THE REAL MARS

By the same author
The Worlds of Galileo

THE REAL MARS

Michael Hanlon

CARROLL & GRAF PUBLISHERS
New York

Jacket imagery

(*Front jacket*) Ophir and Candor Chasma, Valles Marineris, view to southwest
(1/128 degree MOLA data), © Kees Veenenbos

(*Back jacket, main picture*) Valles Marineris detail, HRSC image from Mars Express,
ESA

(*Back jacket, middle inset*) Summit caldera of Albor Tholus, Elysium region, HRSC
image from Mars Express, ESA/DLR

(*Back jacket*, *left inset*) Future explorers locate old Viking lander, NASA

(*Back jacket*, *right inset*) Mars Exploration Rover Opportunity, looking back at
Eagle Crater and discarded airbags, NASA/JPL/Cornell

(*Previous pages*) The interior of a crater nicknamed 'Bonneville' dominates this
false-color mosaic of images taken by Spirit's panoramic camera.

Carroll & Graf Publishers
an imprint of Avalon Publishing Group, Inc.
245 West 17th Street, 11th Floor
New York, NY 10011-5300
www.carrollandgraf.com

First published in the UK in 2004 by Constable,
An imprint of Constable & Robinson Ltd.

First Carroll & Graf edition published in the USA in 2004

ISBN 0-7867-1413-1

Printed and Bound in China

Contents

Acknowledgments

For a non-scientist, attempting a book like this is a perilous exercise. So many facts, so many statistics, so many possible errors of interpretation. I can only hope that I have, for the most part, got it right. If so, this is thanks to those far more expert than me who have given their valuable time to steer me back on track when I was on the wrong path.

I must thank first the staff at the various NASA offices, who have as always been kind and helpful. I thank in particular Jim Garvin, who gave me much of his valuable time at what must have been a very busy period and who kindly agreed to write the Foreword. I must thank David Harland, David Portree, Vic Baker, John Connolly, Monica Grady, Rosaly Lopes, and Nick Hoffman for their insights, help, and advice. My gratitude goes to the Pillingers, who were extremely helpful at a very trying time; and to the staff at the UK's Particle Physics and Astronomy Research Council, especially Peter Barratt, who was on call twenty-four hours a day through the whole Beagle saga. Alistair Scott at Astrium was very helpful, as were the staff of the Lowell Observatory, especially Antoinette Beiser, and the staff at the various journals from which I have quoted in the book. Where published sources are not specifically indicated in the book, quoted extracts are from press releases, verbal presentations, or personal discussions.

I must thank Kate Mayger at the *Daily Mail* for her invaluable help with picture research, and Dick Taylor of the British Interplanetary Society, who was always on hand to offer help and advice. Finally, my thanks go collectively to my publisher, Constable & Robinson – and in particular to my editor there, Pete Duncan – and to my copy-editor, John Woodruff, for wading through my text, correcting a multitude of sins as he did so. And lastly, my infinite thanks to Elena Seymenliyska for her endless enthusiasm and tolerance.

Foreword

by Dr Jim Garvin

Mars has long been so much more than the fourth planet, evoking personal responses that range from cries of war to the most imaginative works of science fiction and film. To some it may seem but a small, rocky planet in a tiny solar system, within a small part of a very ordinary galaxy. But to 'Mars-philes' it is and always will be the very Canaan of the religious passions that embody true exploration. Today we are witnessing the birth of a new Mars, thanks to the vicarious voyages of a fleet of robotic spacecraft that are now uncovering new mysteries and showing humanity that the real Mars is even more intriguing than the planet painted in the stories of sci-fi and modern film.

A passion for exploring Mars as a new scientific frontier has been a part of my life for as long as I can remember. So much so that on occasion I am described as 'one of the real Martians'. Yet perhaps, as Ray Bradbury suggested in his *Martian Chronicles*, we are all the Martians – as we ponder the mysteries of our red-hued planetary neighbor. As people we are drawn to mysteries that offer clues to who we are and where we have come from, and these questions clearly apply to the present wave of exploration of Mars. Perhaps there is hope of making headway in our quest to understand whether we are indeed alone in an infinite universe by searching for life, in whatever form, on tangible, Earth-like worlds such as Mars. So, for me, as for author Michael Hanlon, Mars is much more than one of the inner planets in our solar system; it is a truly special place in the mind's eye.

Now, at the dawn of a new millennium, it is especially fitting to step back and reflect on what Mars has become. Hanlon's perspective on the real Mars captures the highlights of emerging views of the red planet, from the standpoint of a real enthusiast,

yet with the sincerity of someone objectively evaluating a prized possession. While it may be a story that has no ending, it deserves to be told now, as discoveries rock the foundations of current wisdom about what Mars may ultimately tell us about ourselves and our own destiny.

It is humbling to realize how far we have come in the past decade, as Mars has awakened from the slowly changing, inhospitable world that many scientists concluded after the provocative results of the Viking mission in the late 1970s. It has emerged as the best place we know of to aggressively search for evidence of past or present life beyond our home planet. Indeed, the possibilities of life on Mars have catalyzed whole new disciplines such as astrobiology, and promoted an armada of robotic spacecraft to journey to the Martian system to lay ground-work for a relentless campaign to understand the biological potential of another world.

That we, the people of planet Earth, have wholeheartedly embraced this curiosity-driven blend of exploration is encouraging, given its outward-looking character and the purely inspirational and educational value to future generations. This is Mars's time to come alive and shed its secrets so that the urge to go visit ourselves will finally outweigh both the risks and costs of doing so. It has been said that great moments in history require catalytic events driven by human needs associated with power, wealth and so forth, but humanity's current exploration of the real Mars is motivated by curiosity and the possibility that we are not alone.

The stunning discoveries of the Mars Global Surveyor, now beyond its 25,000th orbit of the planet, have painted for us a Mars of new possibilities in which the role of liquid water and its perseverance bodes well for at least the tentative chemical antecedents of life. Today, with the discovery of scientifically

viii

compelling evidence for shallow, salty bodies of liquid water on Mars, thanks to the observations of NASA's rover Opportunity, the real Mars is shaping up as a place less bizarre and enigmatic, yet one in which we may come to learn about how life could have started here on Earth. If Mars could offer us those insights over the next few decades, the value of our voyages of exploration would defy the critics, and pave the way for a renaissance in human space travel.

This book is a personal voyage by a bonafide fan of Mars, capturing the excitement of this unique moment in time as five robotic spacecraft struggle to read the fragmental and beguiling records of another world. As someone who once offered to journey to Mars 'one way', inspired by the activities of the Viking missions of the late 1970s, I can brashly state that the real Mars has always been better than any version of Mars imagined in fiction, film, or even by the media. Mars may be our destiny and *The Real Mars* is a fitting reflection on where we are today as we dream of those future voyages in which the first Earthlings will visit another world as beautiful in its own way as our own. This is indeed Mars's time!

Dr Jim Garvin,
NASA Lead Scientist for Mars Exploration,
2004

(*Opposite*) Rendering, by Kees Veenenbos, of the Gusev Crater in today's conditions, the target for Spirit in 2004.

Author's Preface

Mars Hill, Flagstaff, Arizona; 1 a.m., 29 July 2003

Down in the desert it is still 30 degrees, and people are turning up their air-conditioners so they can get to sleep. But up here, high on a hill a couple of kilometers out from Flagstaff, it is cool, cold even. I wish I had a jumper, or even a blanket. I am sitting on amphitheater-like steps inside a large wooden dome, looking at one of the most influential pieces of scientific equipment in history, awaiting my turn.

Bedecked with engraved brass dials and focusing wheels, it looks more like a giant Victorian steam engine or the innards of a great ship than the precision instrument it is.

This is the Clark telescope, star attraction of the Lowell Observatory. As big as a bus, and with a lens as wide as a coffee table, this impressive piece of nineteenth-century engineering was commissioned in the 1890s from the acclaimed telescope-makers Alvan Clark & Sons by the Boston millionaire Percival Lowell to further his studies of the planet Mars. What Lowell saw through this extraordinary telescope defined the way we think about the most Earthlike planet in the solar system, and continues to define it to this day.

Sitting proudly atop its five-tonne cast-iron plinth, the Clark Telescope is pointing at a brilliant orange dot in the eastern sky. Tonight, the Red Planet is just about as close to Earth as it ever gets; in a month's time the two planets will be closer than they have been at any time since the Neanderthals looked into the skies, just 55,758,006 km (34,646,418 miles) apart. The last time Earth was this close to the fourth planet was in the year 57,537 BC; I would have to wait until 28 August 2287 to get a better look. I am keen to take my place at the eyepiece. But I have to wait my turn, for squinting through the telescope right now is one Richard Hoagland. But more of him later.

There are bigger telescopes in the world than the Clark, of course. This instrument is now used solely for public outreach

Percival Lowell seated at the Clark Telescope at Flagstaff. When it was built it was one of the finest instruments of its type in the world. It is still an excellent telescope, and every year it allows hundreds of children and curious adults the chance to peer at the planets and stars.

purposes. Paying punters and their children can peer through it, and *ooh* and *aah* at the rings of Saturn and the cloud belts of Jupiter. There are bigger and better telescopes through which to view Mars than this venerable device. The great twin Keck Telescopes, for example, perched on their lofty aerie in Hawaii, would give me a far superior view of the planet, as would a hundred other more modern and more powerful instruments scattered across the globe. The Clark hasn't been used for serious scientific research for more than two decades.

But I am not here for scientific research; this is a pilgrimage. And besides, I would have as much chance of being allowed to peer through one of the Kecks as I would through the Hubble Space Telescope, 600 km over our heads. The Clark telescope is famous for one thing: it is the instrument with which the man who commissioned it, Percival Lowell, discovered the canals of Mars. Single-handedly, Lowell created a whole mythology – a whole, virtual, planet. A mythology of an alien world teeming with life, a dying world forced to rely on the technical ingenuity of its inhabitants, who had constructed vast irrigation channels to bring water from the polar caps to the arid equatorial lands. Thanks largely to Lowell's observations, Mars, for nearly seventy years, became a land of mystery and intrigue. It was populated by the belligerent invaders of H.G. Wells and the ethereal ghosts of Ray Bradbury – princesses and monsters, fire balloons and tripods. All of this came about as a result, not of some epic voyage of discovery, but by one man with a keen eye peering through a metal tube.

Mars is an exciting place, and I am anxious to view it, close up. I am keen to see what Percival Lowell had seen. Eventually, my turn comes at the helm of the great telescope. And there it is! The images is shimmering at first, but soon my untrained eye becomes used to it. For three nights the skies over Flagstaff have been cloudy. The moist air of the summer 'monsoon' can give good seeing (the term observers use for the clarity of the atmosphere) when the air is clear, but sadly it gives zero seeing when clouds are around. But there are no clouds tonight – no clouds, anyway, between the telescope I am crooked around and the dusty sands of Mars.

I had been warned not to expect too much by a man who knows more about Mars than most people, the Yorkshireman Michael Carr, who for more than thirty years has studied the planet at the US Geological Survey just down the road from this antique observatory. Carr, like Lowell in his day, has become one of the pivotal figures of Mars. His book on the possibilities of water on the planet is widely regarded. 'When you take a look at Mars through that telescope,' he told me a few days before, 'you wonder how in God's name they drew any conclusions whatsoever. They saw waves of darkening, hazes, and lilac rings. The thing is, the image is so damn small.'

Carr was right. Mars, even at a magnification of 200 through this magnificent telescope, presents a puny disk to the viewer. It looks maybe a quarter the size of the full Moon seen without the aid of lenses – maybe the size of a dime held at arm's length. For several seconds, the muscles in my eye struggle to get my pupil to the right size, and my eyeball strives to focus. But slowly, Mars resolves itself.

And true enough, through the Clark, Mars is a magnificent if rather confusing sight. But even I can see the polar caps. Slowly, other features begin to make themselves visible. I see a large greenish vee-shaped area on the left limb of the planet. This sharpens into a definite outline, extending across the disk as a wave of darkness running to the north. A big dark blob is visible at the bottom right of the disk – could this be Syrtis Major, once known as the Hourglass Sea – the prominent feature whose passage across the planet's disk told astronomers that Mars has a twenty-four-hour day, just like the Earth? To the far right there is a streak of darkness, which in a brief moment of clarity resolves into a clearly delineated structure, like a gash across the Martian surface. It is unlikely, but could I be glimpsing the Grand Canyon of Mars, the series of giant channels known collectively as the Valles Marineris? You are not supposed to be able to see this through a telescope as small as the Clark, but who knows? *Something* is there. Other, better eyes press to the lens, and there is agreement that something dark and narrow is there.

I see no violet haze, but I do glimpse a dark ring around the south polar ice cap – and the suggestions of a ring around the much smaller northern cap, which had just resolved itself into view. The fact that I can see anything at all is a tribute to the workmanship and craftsmanship that produced this lovely old telescope, and to the celestial mechanics that had brought my home planet and the object of my study closer than they had been for nearly 600 centuries.

I see the ice, the rings, and the dark patches, but that is it. No canals, no oases. Maybe we saw that canyon – the only 'canal' that can be possibly seen from Earth. But probably not.

Few people see it even when they know what they are looking for, which I certainly didn't. I squint and press my eye against the glass once more, hoping to see Lowell's Mars flash into view. I try unfocusing my eyes, turning my mind's eye to the side, hoping that I can fool my brain into seeing the hidden detail that remains tantalizingly out of reach. The fact that I know perfectly well that the canals aren't there, that Percival Lowell had fallen victim to eyestrain, relentless optimism, and optical illusion does not quench my disappointment. I had hoped to see Lowell's Mars, through Lowell's telescope, and it was not there. This was not the Mars of canals and princesses. I was seeing another Mars – the real Mars.

Why Mars? It is a good question. Why not write a book about Venus or Neptune? Go to any astronomy bookshop or website or magazine and you will find books, articles, and papers about the Red Planet seem to outnumber those about all the other objects in the solar system. The answer is simple: Mars is the most Earthlike place we have seen in space. It is the only object beyond the Moon whose surface is directly and easily observable from Earth. This is because it is relatively close and has a thin, transparent atmosphere, enabling the owner of even a small telescope to pick out features that have fascinated and puzzled astronomers for centuries.

Mars is an important planet to us and always has been. Its blood-red color suggested violence to the ancients. Although the planet's warlike connotations were never in doubt, Mars did play a secondary role as a farmer, the god of agriculture. Roman warriors made iron sacred to Mars – they would have been pleased to discover that metal features strongly in the modern scientific quest to understand this planet – and often had amulets made from iron to protect them in battle.

Mars is smaller than the Earth, with lower gravity and a thin, frigid, poisonous atmosphere. Yet compared to anywhere else in the solar system, Mars is a veritable paradise. On most planets you would be boiled, crushed, irradiated, or frozen in an instant; on Mars, death would take minutes rather than milliseconds. Mars has a twenty-four-hour day, four distinct seasons, and a landscape that strongly resembles that of Earth's great deserts. With its red rocks, canyons, and mesas, much of Mars could be the Arizona Desert or the South African Karoo.

Yet Mars is alien. On its surface nothing stirs, save a thin, piercing wind that can hardly be felt yet is capable of whipping dust off the surface in such quantities that the entire planet becomes a muffled, orange ball. The atmosphere is a wisp of carbon dioxide, so tenuous that if exposed to it the blood in your veins would start to boil. The soil is a fine dust, so fine (the particles are ten times smaller than those in talcum powder) that it almost flows like a liquid. Visitors will not only have to cope with the obvious dangers of vacuum and cold, but also with less-obvious hazards such as corrosive soil, a constant rain of ultraviolet light, and an extra thirty-six minutes in the day – which will make astronauts feel progressively jetlagged.

Just how like the Earth – or how unlike the Earth – Mars is, has become one of the great debates in planetary astronomy. As a science journalist I have become fascinated by how this planet seems to polarize scientific opinion like no other. Some scientists believe that Mars is far closer to our own world than we have been led to believe by the first generation of space probes. Photographs from the Mars Global Surveyor orbiter and data returned by the latest robotic landers appear to show a world fashioned, at least in part and in the ancient past, by water. There are gullies, water channels, melting snowpacks, and the ghosts of ancient salt pans. This new Mars truly resembles the harshest places on Earth. Then there are those who accuse the 'warm, wet Mars' brigade of wishful thinking. Mars, they say, is a harsh, hostile planet and always has been. It is too cold, too dry to be anything like Earth. Carbon dioxide, rather than water, is the agent that has carved out the bizarre landforms we see today. Occupying the middle ground, and in the majority, are those who see a Fallen Mars, a once clement world which sank into a deep-frozen hell. All over Mars, these scientists say, are the ghosts of a warmer, wetter past – ancient river valleys, deltas, and shorelines. Exploring Mars is to explore a lost world that has been bypassed by 3 billion years of time.

These views are strongly held and hotly debated. Scientists attack their opponents with bitter sarcasm – these are not esoteric arguments. Mars has polarized the scientific community like few other subjects – and that alone makes it a fascinating subject for a book.

A hundred years ago there were equally heated arguments,

(*Opposite*) The Martian south polar cap in midsummer, photographed by the Mars Orbital Camera aboard Mars Global Surveyor in April 2000. It is probably cold enough for most of the carbon dioxide to remain as dry ice, although the bulk of what you are looking at is currently thought to be water.

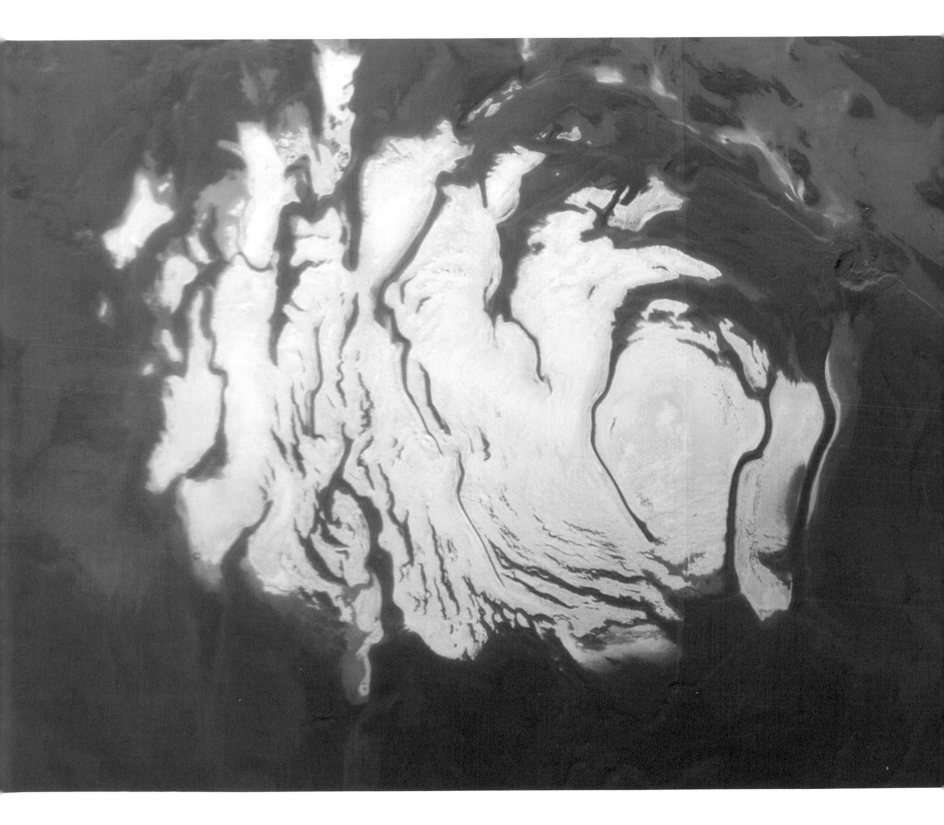

but about the nature and ambitions of the planet's inhabitants. The canal-builders imagined by Percival Lowell had motives and lives we could only guess at. Those who believed Mars was dead were in the minority. Now we know far more about Mars thanks to our spacecraft, but still the arguments rage. The more data we have, in the form of photographs and spectra, the less it seems we know for sure about this enigmatic planet.

In the end, the great debates about Mars all boil down to one question – the question of life. NASA's current strategy is to 'follow the water', not just because water is particularly interesting in itself, but because strong evidence that water flows – or once flowed – on the surface of Mars would be strong circumstantial evidence for another planet capable of sustaining a biosphere. If we do find life on Mars – dead or alive – it will provoke a profound paradigm shift, not only in the world of science but also in the very way we think about ourselves. Life on Mars would have profound implications for philosophy, religion, and the relationship between us and the rest of creation. That is why NASA is investing tens of billions of dollars into the conquest of Mars, and that is why Mars exploration has caught the public imagination like no other space enterprise since the days of Apollo.

As this book was being completed, there were intriguing signs – from the Mars Express orbiting spacecraft and from Earthbound telescopes – that the telltale signs of life had indeed been seen in the Martian atmosphere. If the findings are confirmed, they will totally rewrite the history of the Red Planet.

From the announcement by President George W. Bush that NASA's focus is to shift from low Earth orbit and the much-criticized International Space Station and onto the reconquest of the Moon and an accelerated program to explore Mars it is clear that the Red Planet will be one of the main space targets in the new century. Thirty years ago, Mars was dismissed as another version of the Moon – cold, dead, battered by meteorites. Now Mars has come alive again, its surface giving up clues to ancient rivers and seas.

Many books have been written about Mars in the past couple of centuries. Most have taken the side of those who tend to see Mars as more rather than less like the Earth. This is the Mars of popular imagination, the Mars of Lowell and the canals, of little green men and invaders. Lowell's Mars is long dead, but the Lowellian view persists. I believe that the arguments about what Mars is really like are almost as fascinating as the planet itself. Mars, in a small way, is held up as a mirror to the Earth. What we see in its red disk tells us a lot about ourselves. Whether we are seeing the real Mars – through a telescope, or from close up through the cameras aboard our rovers and orbiters – remains to be seen.

Introduction
Lowell's Shadow

'Are physical forces alone at work there, or has evolution begotten something more complex, something not unakin to what we know on Earth as life? It is in this that lies the peculiar interest of Mars'

Percival Lowell (*Mars*, 1895)

Mars fever, they called it. For the first time in human history, the public's gaze had been directed towards a heavenly body in the name of science. Mars fever – a rash of speculative articles in the newspapers and journals of America and Europe – was at its height exactly 100 years before this book was written, and the fever was largely the work of one man: astronomy's great antihero, Percival Lowell. The Boston millionaire, founder and lifetime director of the observatory that bears his name, single-handedly created the Mars of popular imagination. In the years between the founding of the observatory in 1894 and Lowell's untimely death in 1916, the fever at times reached such a pitch that it was as much a talking point among scientists and the public as Darwin's theory of evolution had been towards the end of the nineteenth century. For two decades, Mars came under what was in effect a telescopic siege: the most prolonged and systematic study of a planet ever undertaken. Lowell and his assistants spent night after night peering at its tiny red disk, and several times his telescopes were dismantled and rebuilt on different continents to get a better view.

A century later, a more measured fever has returned. The elegant disk of Mars once again graces a hundred newspaper front pages, and magazines and newscasts are full of stories of missing space probes, surreal pictures being taken from Mars' orbit, and amazing discoveries by the latest NASA rovers wheeling themselves across the enigmatic surface of the planet. Mars has become the subject of presidential dreams and for the first time in decades there is serious talk at NASA about humankind's manifest destiny to explore the solar system. None of this would be happening if Lowell had not seen those

canals. The father of modern Mars came up with ideas which, even after they were shown to be completely wrong, have for more than a century shaped the way we think about this planet.

The speculation sparked by Lowell's observations, and his interpretations of those observations, concerned the nature of the Martian surface, the evolution of the planet and, most of all, the nature of the beings who might inhabit it. Of course, even then not everyone took Lowell's ideas seriously, but they were, by and large, accorded respect. A hundred years ago there were fierce arguments as to whether the Martian surface was home to water, with some scientists maintaining that it was far too cold and the atmosphere far too thin, and others (the majority viewpoint) maintaining that Lowell and his acolytes had shown beyond all reasonable doubt that a vast network of artificial canals spread across the Martian plains.

Lowell presumed that Mars' advanced state of desertification signified a ubiquitous process that he formalized in his new science of 'planetology'; and the idea took hold that Mars was a dying planet, inhabited by intelligent beings making a last-ditch stand against the encroaching sands with a global engineering project to irrigate the lower latitudes with melted ice from the polar regions. At the time canals were seen as perhaps the ultimate in hi-tech building projects. The Panama Canal was under construction, so it was natural to assume that the advanced Martians would be undertaking similar projects.

By the time he died, Lowell had recorded more than 700 canals, many sprouting in a single month or less. The media had a field day. Martians were a great story, and soon there

were reports about their attempts to get in touch with Earth, as periodic 'projections' were seen extending from the planet's surface. These patches of light were interpreted by some as messages aimed at Earth by the denizens of our 'brother planet'. After Lowell's announcement of such a 'projection' in 1903, the London *Daily Mail* attempted to alarm its readers:

> The brilliant imaginations of H.G. Wells have familiarised us all with the possibilities of life on Mars, and no one who has read *The War of the Worlds* can help shuddering slightly when he remembers that just such a projection indicated the commencement of that terrifying invasion.

To be fair, the article went on to admit that it was 'practically impossible that mankind should ever have to fight for its life against a Martian army', pointing out that conditions on the Red Planet were probably harsher than many were supposing.

Although there were few who advocated travelling to Mars (no one had much of a clue about how this could be accomplished), there were many who speculated on the Martians travelling to *us*. Lowell's Mars became the Mars of fantasy as well as of science. It was Barsoom, the irrigated world of Edgar Rice Burroughs, and the dying world of Bradbury's *The Martian Chronicles*. Few, if any, science fiction writers chose to make Mars an uninhabited desert.

Now Mars is again in the news. Not only has the American President announced plans to send humans to the Red Planet, there is now talk of Martian colonies and trillion-dollar, multi-decade missions to what Mars enthusiasts insist will be humankind's new home. A small armada of space probes has already been dispatched, and in August 2003 the planet shone more brilliantly in the skies of Earth than at any time in the past 60,000 years. Spectacular pictures taken from the Martian surface are being beamed back to Earth from the twin American

(Opposite) A historical sequence of images of Mars, showing both the improvement in observation technologies over time – and changes in how the slowly resolving face of Mars was interpreted. The earliest image (upper right) shows the Mars of the early telescopic era, where only the largest and most distinctive features – the Polar Caps, and the 'Hourglass Sea' could be seen. We then move on to the Mars of Lowell (second from top), resplendent with canals, and the Mars of the space age.

rovers, Spirit and Opportunity, showing rock-strewn plains, layered rock formations that were deposited in water, and rather grand, distant hills. No fewer than three orbiting spacecraft are mapping the surface of Mars, in a variety of wavelengths so as to gather different types of information, from an altitude of just a few hundred kilometers, while more are being readied for the long voyage – how Lowell would have loved such a vantage point from which to view his beloved planet. No longer do we have to rely on hazy, wobbling images and Earthbound spectroscopes to scour the distant sands of Mars. Our robots are prodding and scanning, up close and personal. Twenty years from now, Mars will be home to more than a dozen more pieces of terrestrial hardware.

Mars is the subject still of much frenzied scientific debate. Its surface is covered not with canals but with long, meandering valleys and channels. Every week, it seems, brings another discovery that points to life on Mars. The consensus is that the great channels imaged from orbiting space probes were formed by water, which in Mars's distant past flowed across its surface in prodigious quantities. Some maintain that Mars is a fallen world, a planet thrown off balance aeons ago by some cosmic catastrophe whereupon evolution was stopped in its tracks, and what was once a green-and-blue planet has since been overrun by desert sands to become the orange world we know today. Maybe Mars was once home to oceans the size of the Mediterranean, and rivers longer than the Nile.

Results are already flooding in from the robot geologists trundling across the Martian surface. In the coming years, before humans set foot there, robot excavators will drill and dig and then fly back to Earth with their booty. Is there really life on Mars? The optimists point to mysterious shapes, microscopic tubules glimpsed in a meteorite plucked from the snows of Antarctica and presumed to have been blasted off the surface of Mars in a cosmic collision. Are these the remains of ancient bacteria? Possibly life still thrives there today, kept warm by geothermal heat from ancient, smoldering volcanic chambers. Dig deep enough, say the optimists, and we will find not only water but bugs. The jury is still out on a second Genesis on the fourth planet.

Lowell was attacked throughout his professional career for his elaborate speculations. Today, not everyone is convinced by the consensus that Mars once had lakes and rivers, even life. Some think we are looking at too little and seeing

too much. To the skeptics, Mars is still an enigma, a world superficially like the Earth yet fundamentally so different – a colder, harsher version of its big sister planet where life could never have got started. Maybe those channels were carved not by water but by more exotic, alien substances like liquid carbon dioxide, or even by winds.

There are also those given to more fanciful speculation who see faces, pyramids and cities in the mesas and dunes snapped by the orbiting probes. But just why are humans so fascinated by Mars? What is it about a place that for millennia was merely an orange dot in the sky, albeit one which shines power-fully during its two-yearly close encounters with the Earth? We speak commonly of Martians but far less so of Venusians or Jovians. Mars is vaguely belligerent, hostile. Mars was not only the god of war, but a place from where war was visited upon us, from the unlikely bridgeheads of Dorking in Surrey and Grover's Mill in New Jersey (the targets of Wells's invaders, in the original and in the later American radio versions). It was a world of jealous conquistadors enviously eyeing their blue-green neighbor across the void. Does our fascination for Mars stem from the place itself? Or does it actually say more about us and our desire to not be alone in the Universe?

Percival Lowell believed Mars was most definitely a place he could explore at will without the discomforts and privations of a trip up the Amazon, say, or a hike up the mountains of central Asia. He could see it all from his telescope, situated a handy one-minute walk from his house atop the hill he had purchased outside the little Arizona town of Flagstaff. Following the lead of others who had reported having seen strange linear markings on the surface of Mars, Lowell duly saw not only canals, but great shifting patches of green that for decades were widely assumed to be waxing and waning patches of vegetation, or maybe even gigantic swarms of locusts or mos-quitoes. He saw 'waves of darkening' and the dark blue ring around the poles. His canals, as he continued to observe them,

The Red Planet looming in the skies of its moon Deimos, from where a group of astronauts are admiring the view – as envisioned by space artist Chesley Bonestell. By the 1950s, most astronomers considered Lowell's Mars to be a scientific fantasy, but the canals lingered on in the popular imagination. It took later visits by real space probes to erase them from the surface of Mars.

became ever more complex. Single channels often became double, or 'geminated' to use Lowell's preferred term. Where the canals intersected, dark, triangular patches could be seen. Perhaps these were cities, or desert oases.

To get a feel for what Lowell saw when he looked through his magnificent telescope, here is an extract of a report he filed for the Associated Press on 2 January 1916:

> Results of signal importance have already been brought out in that the observed canal development is strikingly corroborative of the theory of seasonal dependence of the melting of the polar cap. The northern canals are now very dark indicating increased activity with advancing spring, while the southern canals are faint in their autumnal decline. The season in the northern hemisphere of Mars is now late April.

Note the certainty of Lowell's words. He was seeing not just a *feature*, a faint patchwork of lines on the Martian surface that he was interpreting as a network of canals. Instead, he was seeing a *story*, writ large – or rather, writ small – on the tiny, shimmering orange disk he saw through his eyepiece. He imagined a huge thaw of the polar caps that came with spring, trillions of gallons of blue water gushing, in this case south, from its icy winter holdfast. There was vegetation – mighty trees, perhaps, or swathes of shrubbery.

Lowell believed that Mars was home to intelligent life – not human life, as he was at pains to point out, but nevertheless rational, thinking beings advanced enough to be doing civil engineering on a planetary scale. He wrote countless letters and articles and three books expounding his theories, and gave hundreds of lectures which were nearly always sellouts. Lowell was engaging and highly articulate, both in print and in person. He had a worldwide fanbase and received letters every day from scientists and from members of the public offering their own ideas about the Martians and requesting more information from him. He was probably the most famous astronomer of the modern era.

Percival Lowell knew that to be visible from Earth, the 'canals' had to be a least 50 km (30 miles) across – an infeasibly large structure. He reasoned that what we were seeing were wide strips of cultivated land either side of the waterways – ribbons of green much like the verdant strips that run alongside the River Nile here on Earth.

The man who started it all. Percival Lowell, a Bostonian millionaire, set up the observatory that bears his name on a hill outside the little town of Flagstaff in Arizona. Although the canals he saw were not real, his thesis – that Mars is a planet much like the Earth – has inspired more than a century of observation and exploration.

As late as the 1950s, the consensus was that Mars was home to life, perhaps advanced, multicellular life. Maybe the 'violet haze' in the Martian atmosphere, reported by many astronomers, was some sort of protective ozone layer, allowing vegetation and fauna to flourish. If there *was* ozone in the Martian atmosphere, then there was 'ordinary' oxygen also, and there was life to generate it because an atmosphere with oxygen cannot be stable. When it became clear that the canals would have to be at least 50 km across to be visible through

telescopes, Lowell refined his ideas. These were not the canals themselves we were seeing; instead, the dark lines that criss-crossed the surface, following precise great circles and looking like a map of international air routes, were the visible man-ifestations of linear civilizations, Martian Niles which, like the African river, are visible from space as ribbons of green in sharp contrast to the orange desert sands.

Lowell even thought he had gained some insight into the minds of the Martians. He deduced that the 'Martial mind' was mathematical, comprehensive, inventive, cosmopolitan, and technically superior to our own. Lowell, a deeply conservative man yet something of a pacifist, assumed that his Martians had put their differences aside and embarked on a global engi-neering project for the good of their species as a whole. Like Wells and many science fiction writers since, Lowell saw his Martians as a rebuke to human vanity. For while Darwin had, a few decades before, knocked man from his pinnacle as the terrestrial embodiment of God, the Victorians at least had the compensation of learning that *Homo sapiens*, if not divinely created, was at least at the top of Darwin's evolu-tionary tree. The fact that this ancient world, Mars, was home to creatures capable of such feats of engineering suggested that not only was humankind not alone, but that we had most likely been far outstripped in our achievements.

Lowell's Mars was a home of intelligent life-forms struggling to save their dying cities from the ravages of drought. Mars's red sands became a symbol of age, a timeless desert upon which ancient civilizations had risen then fallen. Where science led, the imagination followed. If Lowell's and Wells's Mars was Ancient Greece, the Mars of pulp sci-fi was Egypt or Mesopotamia, where the sands slowly smothered the remains of ancient wonders.

Some skeptics doubted that Mars was quite as Earthlike as popular imagination demanded. As early as 1898, Cowper Ranyard and George Johnstone Stoney in Britain, eschewing the tempting analogy with the Earth's polar caps, argued that the Martian ice caps were a frost of frozen carbon dioxide. The presence of dry ice at the Martian poles suggests an alien meteorology quite incompatible with Burroughs' scantily clad princesses or even tough warriors. Alfred Russel Wallace, Darwin's fellow evolutionist, was unimpressed by Lowell's views of Mars as a home to a great civilization of canal-builders. 'Not only is Mars not inhabited by intelligent beings, but it is uninhabitable', he wrote. There is no evidence that Wallace

'Not only is Mars not inhabited by intelligent beings, but it is uninhabitable'

Alfred Russel Wallace

ever actually looked at Mars through a telescope but, interestingly, even he accepted Lowell's assertions that the surface of that planet was crisscrossed by a giant network of linear features – features he attributed to volcanic activity and planetary shrinkage. He wrote:

> All physicists are agreed that, owing to the distance of Mars from the Sun, it would have a mean temperature of about −35°C ... if it has an atmosphere as dense as ours ... But the very low temperatures on Earth at the Equator, a height where the barometer stands about three times as high as on Mars, prove that ... Mars cannot possibly have a temperature as high as the freezing [point] of water ... the temperature of Mars [is] wholly incompatible with the existence of animal life.

Lowell reacted with fury to Wallace's rubbishing of his ideas. In a letter to *Nature* published on 6 March 1908, he sniffed that Wallace is clearly 'expecting some sort of reply'. He continued:

> The effect of its perusal is to show me again how cogent is the argument for Mars' habitability, for only by many misstatements of fact, wholly misunderstood of course, can Mr. Wallace make out even a plausible case upon the other side.

Lowell then launched into a criticism of how Wallace must have miscalculated the amount of water present in the polar caps:

> An equally fatal flaw affects Mr. Wallace's argument for temperature ... such errors cannot be too carefully avoided in science, especially when a man, however eminent he is in one branch, is wandering into another not his own – indeed, if criticisms are confined, as commonsense wisdom counsels, to those versed in the phenomena, we should hear very little of Mars's inhabitability.

The British historian of astronomy Agnes Clarke, writing in the *Edinburgh Review* (1896), was equally critical. She pointed out that Lowell's canal network seemed to defy the laws of

topography: 'We can hardly imagine so shrewd a people as the irrigators of Thule and Hellas wasting labour and life-giving fluid after so unprofitable a fashion.' One anonymous journalist summed up the essence of skepticism about Lowellian Mars thus: 'Prof. Percival Lowell is certain that the canals are artificial. And nobody can contradict him.' Two of the staunchest opponents of Lowell's theories were the astronomers Edward Holden and Wallace Campbell of the Lick Observatory. Campbell, who had failed to detect any evidence for water in the Martian atmosphere, commented:

> In my opinion [Lowell] has taken the most popular side of the most popular scientific question about. The world at large is anxious for the discovery of intelligent life on Mars, and every advocate gets an instant and huge audience.

Nothing, of course, has changed. A hundred years later, silly stories about Martian face-carvers and even sensible stories about Martian microbes get onto the front pages, whereas stories about Martian sand dunes and volcanoes struggle much harder to make the news.

Some people decided to put Lowell's ideas to the test. After his telescopic observations during the 1903 opposition (the

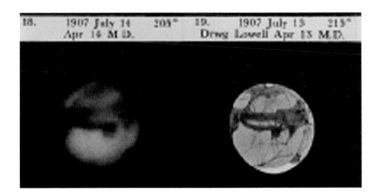

A drawing of the planet Mars and its canals made by Percival Lowell in July 1907, and a photographic image of 'his' planet taken a day later through the same telescope he was using when he made the drawing. Lowell hoped that telescopic photography would settle once and for all the question of whether the canals were real; in reality, the small, blurred images were no match for hundreds of hours of telescopic observation by the human eye and interpretation by the human brain. The photographs simply added fuel to the fire of the debate. Just as through the eyepiece, some saw canals in the photographs, while others did not.

time when the orbits of Mars and the Earth bring the two planets the closest together), Lowell became even more convinced of the reality of his canals. But although he found widespread support in America, across the Atlantic things were different. British science was generally more skeptical, and in June of that year the London astronomer Walter Maunder reported to the Royal Astronomical Society that when a group of boys from the Royal Hospital School at Greenwich were asked to copy an image of Mars, showing only the light and dark patches but no canals, they invariably drew a planet crisscrossed with lines. He wrote:

> It would not therefore be in the least correct to say that the numerous observers who have drawn canals on Mars during the last 25 years have drawn what they did not see ... On the contrary, they have drawn, and drawn truthfully, that which they saw; yet, for all that, the canals which they have drawn have no more objective existence than those which our Greenwich boys imagined they saw on the drawings submitted to them.

Maunder's result was dismissed contemptuously by Lowell as the 'small boy theory', but the idea that the canals were simply a mirage – an attempt by the brain to make sense of the visual 'white noise' that is the mottled, meaningless image swimming in the telescope eyepiece – began to take hold. Other astronomers, most notably Eugène Antoniadi, originally a supporter of Lowell's ideas, began to have doubts. By 1909, the French astronomer, who had access to a magnificent 33-inch (0.84 m) telescope in Paris, larger than Lowell's, was able to declare that 'the geometrical canal network is an optical illusion'.

Lowell's assistant Vesto M. Slipher used newly acquired infrared-sensitive film to make a spectroscopic analysis of the Martian atmosphere. In 1908 he found evidence that the atmosphere was rich in water vapor. After Slipher wrote to his boss, Lowell announced to the world that his team had obtained definitive proof of the existence of water on Mars, and a report was published in the journal *Nature*. But by now the press was becoming more wary about Lowell's Martians. The *New York Times*, a friend of the astronomer's ideas, published a piece about the findings of water-on-Mars, headlined 'A STEAMHEATED PLANET'.

But, as telescopes grew bigger, they started to suck the

The Lowell Observatory is about 2,200 m (7,250 ft) above sea level. Lowell, Slipher and the other astronomers who lived and worked there enjoyed clear skies and relatively little light pollution; but they also often experienced heavy winters, much different than the weather on the desert plains far below – a price worth paying for the clear thin air.

water out of the Martian canals. When George Ellery Hale observed the planet with the Mt Wilson 60-inch (1.5 m) reflector, then the most powerful telescope in the world and vastly better than Lowell's 24-inch (0.6 m) refractor at Flagstaff, he saw 'no trace of narrow straight lines or geometrical structures'. The main problem for Lowell and his adherents was that photographing the canals was proving to be impossible. Even with the best cameras and Lowell's finest lenses, the 'canals' which he thought he saw steadfastly refused to show themselves on the photographic plates. In November 1903, Lowell remarked to his assistant Slipher that 'It would be encouraging to have an undisputed canal show in a plate.' Slipher dutifully began taking pictures.

These turn-of-the-century doubters were voices in the wilderness, however. The consensus was that Lowell was on to something, even if he was finding difficulty proving that his canals were real. Throughout the twentieth century, although it became

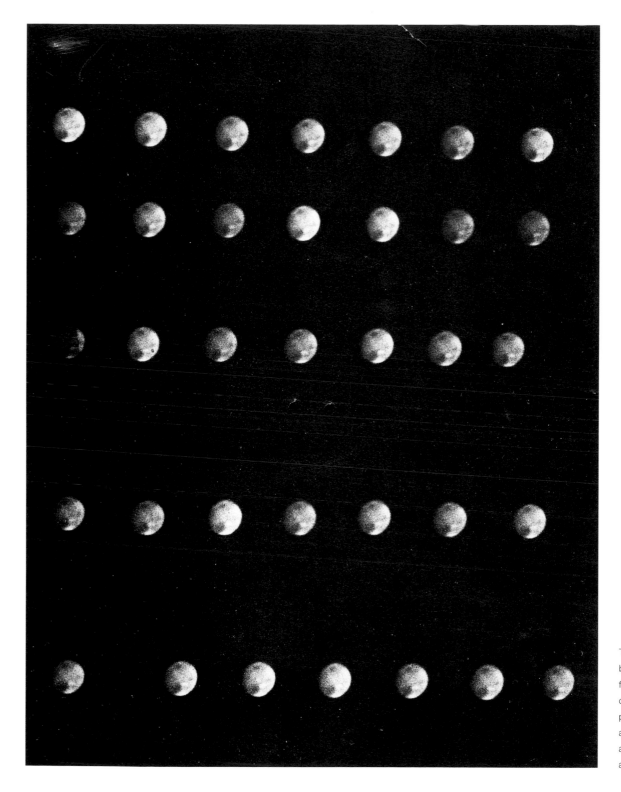

These photographs of Mars were taken at the behest of Percival Lowell, who was keen to find evidence of his precious canals that would convince everyone. When a series of Mars photos were put on public display, they caused a sensation, although many visitors were disappointed that the disks were so small and appeared not to show any canals.

clearer and clearer that the Mars of Lowell was probably a fantasy (it was simply too cold and its atmosphere too thin), the vision of Mars as a kind of alternative Earth never quite went away.

'We must admit that observers have too often begun the study of Mars at the wrong end'

Gerard de Vaucouleurs

The French-American astronomer Gerard de Vaucouleurs wrote in 1939 that 'Everybody has heard of the "canals" of Mars; these strange appearances are described by some people as proof of the industrious activity of "Martians", yet others will not admit that they are anything but persistent optical illusions.' He quoted the director of the Paris Observatory, André Danjon: 'It is not by nature that an object of study is or is not scientific, but only by the manner of looking at it.' To which Vaucouleurs added, 'We must admit that observers have too often begun the study of Mars at the wrong end [of their telescopes].'

In 1939, astronomers were still limited by the technology of ground-based telescopes – telescopes which did not have the advantages of computerized correction for atmospheric disturbances bestowed on today's giant machines. De Vaucouleurs mentioned the 'dark ring' that accompanies the polar caps as they shrink. This suggests, he said, the existence if not of a polar sea, 'then at least of a zone several hundred miles in width where the soil is softened and dampened by the melting ice.' But he admitted that there are problems with this explanation. For a start, the band appears to be of constant width. If it really was a circumpolar lake or a ring of waterlogged ground, then by the laws of perspective it should appear wider to the right and left of the cap in question than at the meridian. This alone suggested that there is something illusory about the band. What was more,

> it is scarcely visible on photographs, suggesting – if the photographs are at all comparable in clarity with visual observation – that the fringe is illusory and arises simply from the subjective effect of contrast due to its close proximity to the sparkling polar whiteness.

De Vaucouleurs was convinced that the Martian polar caps were in fact very thin, 'in no way comparable to the thick permanent beds of the ice fields on Earth.' How did he know this? Calculations of the speed at which the temporary ice disappears in the spring led to the conclusion that the caps cannot be more than '0.1 to 10 inches' (2.5–250 mm) thick. Of course, he was not considering that the caps might be composed of a substance other than water.

It is interesting how many firm conclusions were drawn by the mid-twentieth century that we now know to be false. Forget the always-speculative canals. In 1939, De Vaucouleurs was able to state with confidence, after a study of the planet's disk, that 'If there are any mountains on Mars they can scarcely exceed five or six thousand feet [1,500–1,800 m] in height, and must be more like ancient plateaux than well-marked chains in sharp relief.' He was right about the second point, but way off the mark with the first: Mars certainly does have mountains, and some quite impressive ones too – Olympus Mons exceeds his lowly plateaux by a factor of over twelve. In fact, it is the largest volcanic edifice in the solar system.

When it came to life on Mars, De Vaucouleurs criticized the fashionable view that the 'waves of darkening' were caused by waves of vegetation. If the expanding dark-greenish patches on Mars were plants, and the green color was chlorophyll, then how could that be, considering that there was little atmosphere, and hence chlorophyll would be unable to harness the energy of sunlight to initiate the conversion of atmospheric gases and water into sugar? He concluded that the 'canal question' was far from settled – and indeed seems to have leant towards the skeptical – but admitted that 'we cannot doubt that there is "something" remarkable on Mars'.

The British astronomer and author Patrick Moore has written a series of informative and highly entertaining books on Mars, updated every decade or so, beginning in the 1950s. Each edition has provided a snapshot of current thinking about our neighbor. The 1956 edition contains the following quote:

> The dark areas of Mars are made up of something that lives and grows – it is a cheering thought. Though Venus may be covered with a lifeless ocean, although Mercury and the Moon may be inert, and although the outer planets may be chill globes with poisonous atmospheres, it is pleasant to learn that we are not utterly alone in this solar system of ours.

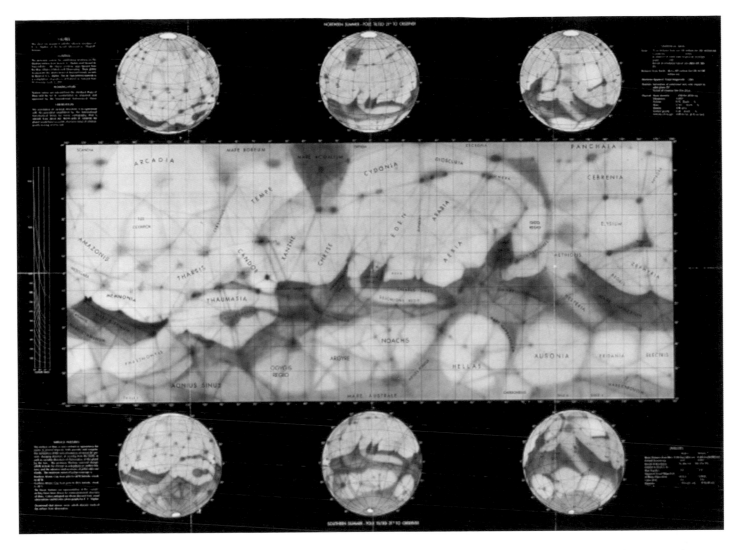

Although Mariner 4 blew away the idea of canals on Mars, the notion never quite died away in some people's minds. This map of Mars was produced, by the US Air Force, in 1967 – after Mariner 4 had sent back its data. The canals (or at least their ghostly imprints) are still there!

As late as the 1960s, most planetary astronomers still assumed that Mars was probably home to lichens, or some similar primitive form of life, although all hope of finding canal-builders was disappearing fast. It is extraordinary now, reading those accounts, to see how readily, well into the late twentieth century, the idea of Martian life was accepted.

In fact, as the space age dawned, and despite the emergence of the new, 'scientific' Mars as a land of cold and almost airless deserts, the canals still lurked, like a glistening, ghostly blue spider's web across a dead world. As a child, I remember reading *Kings of Space* by Captain W.E. Johns, most famous for the Biggles series. Written for a young audience, this was a yarn about an eccentric Scottish inventor who built himself a flying saucer. The adventurers within traveled to Venus, where they found the regulation Carboniferous swamp, complete with oddly transplanted brontosaurs, and thence to Mars. There they found, of course, canals, their banks lined with deserted, crumbling cities, and the whole surface of the dying planet plagued by swarms of mosquitoes – the source of the seasonally varying dark areas on the surface. This book was published in 1954, and was scientifically out of date even then. But for Captain Johns the canals had just refused to vanish into the sands of the real Mars.

They were still there too for Lowell's great champion, Earl C. Slipher (Vesto's brother), who as late as 1962 in his book *Mars, the Photographic Story*, wrote of the unique canal-oasis network system of Mars. This wonderful Martian network has no counterpart on any other planet and is unmatched by any features in nature. Because of the artificial character of the markings they have become the most mysterious and yet the most discussed and most widely debated of all planetary features.

Then, in July 1965, came Mariner 4, which blew apart Lowell's vision. The twenty-one pictures sent back by this probe almost killed the Red Planet overnight. Mariner 4's Mars looked like the Moon – craters, craters everywhere, and not a drop of water to be seen. The huge number of these circular scars imaged by Mariner 4 implied a Martian surface that had not

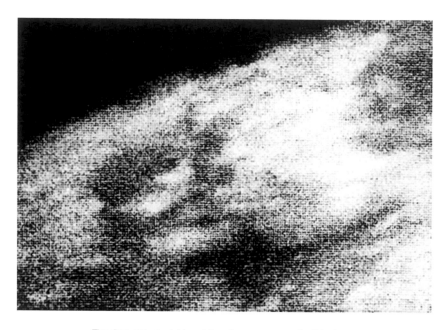

The first picture of Mars taken from a spacecraft. This image was snapped by Mariner 4 as it flew past Mars on 14 July 1965. The pictures, played back from a small tape recorder over a long period, showed lunar-type impact craters (which at the time were just beginning to be photographed at close range by lunar probes), some of them touched with frost in the chill Martian evening. Mariner 4 was only supposed to last for only a few weeks, but it actually survived for three years, generating much useful data while in Solar orbit.

(*Opposite*) Mariner 9 went some way towards redeeming Mars in the eye of many beholders by showing that much of the surface was covered by strange channels, mighty canyons, and titanic volcanoes. This image shows fractured terrain near the Valles Marineris complete with channel-like features.

changed for aeons. They were interpreted as ancient impact craters, although some people argued for years afterwards that they were of volcanic origin. The probe also measured the atmosphere and found it as wanting as the geology. Optimists had held out for a surface pressure of 20–50 millibars – an atmosphere too thin to breathe (or, importantly, to even support breathing), but still thick enough to allow water to flow at reasonable temperatures. Bad news: the Martian atmosphere was painfully tenuous, maybe 7 millibars at best (we now know it can be a bit thicker than this, but not by much). This was not what those who wanted Mars to be like the Earth wanted to hear. Seven millibars is perilously close to the triple-point of water, the pressure at which liquid H_2O is not stable, and any water splashed onto the surface would boil and freeze simultaneously. Lots of craters, an ancient surface, no sign of water erosion, no sign of tectonic forces. And most importantly, no air. Mars was, if not dead, at least on critical life-support.

It is possible that humankind's interest in Mars might have waned completely after the 'disappointments' of Mariner 4. But even now, Lowell's canals could still work their magic. As Richard Taylor of the British Interplanetary Society, says, 'After Mariner 4 it was touch and go whether there would be a Mars program at all, but there was – because of Lowell there was an emotional attachment to life on Mars, so the program continued.' The next two probes to Mars – Mariners 6 and 7 – sent back more gloomy data: thin air and frigid temperatures.

But there was life in the old Red Planet yet. Mars started to twitch with Mariner 9, which arrived in orbit in November 1971. The probe turned up when, Mars was wrapped in an all-enveloping dust storm, shrouding the wonders of its surface from view, much to the chagrin of the Mariner team. But the dust soon lifted, and what delights there were to be seen. Mariner 9 gave us our first proper, close-up view of Mars, and astronomers soon realized that the cratered terrains seen by the earlier spacecraft were not typical. Mariner 9 saw not only scars but immense volcanoes, huge depressions, and, most importantly, immense canyon systems and what looked for all the world like dried-up riverbeds. One gigantic network of canyons – the Valles Marineris (the Valleys of the Mariner, named in honor of their robotic discoverer) – is longer than the United States is wide, and four times the depth of the Grand Canyon.

But it was the 'riverbeds' that drew the most attention. Some, such as the feature named Vallis Kasei, are hundreds of kilo-

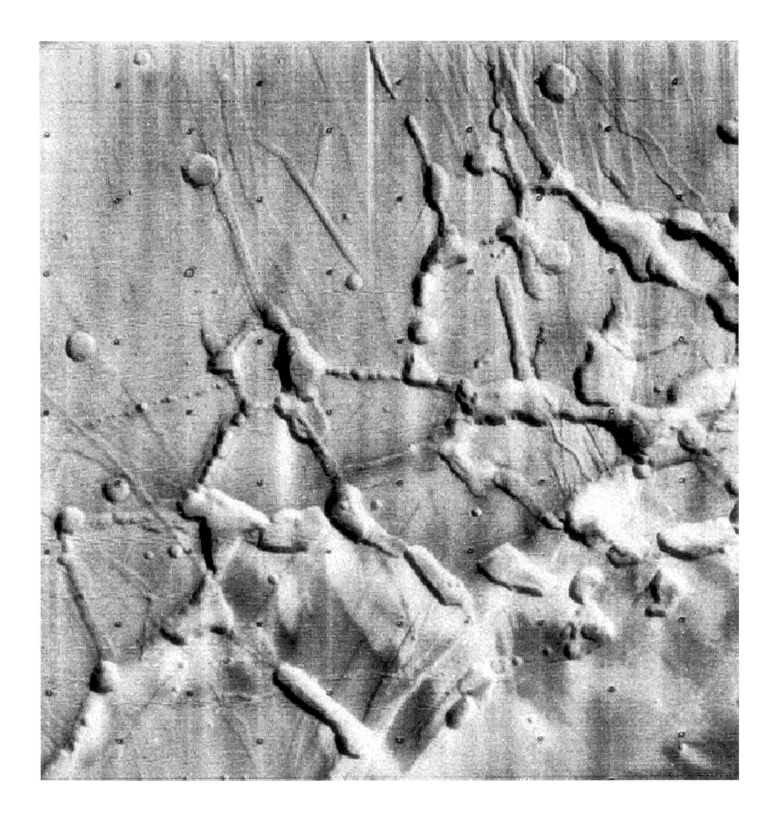

meters long. The Martian surface is covered with hundreds of these valleys. These alleged riverbeds are not, of course, Percival Lowell's canals. Few correspond even vaguely to his misty-eyed observations (and all are far too small to be seen in nineteenth-century Earthbound telescopes), although the Valles Marineris – a network of connected canyons, and the largest linear feature on Mars – does line up with one of Lowell's imaginings (the canal named Coprates– the name of which has now been assigned to a section of Mars's 'Grand Canyon'). And they were, of course, completely dry. Mars is too cold, and its atmosphere too thin, for water to flow in any quantity on its surface today. The real discovery of Mariner 9 was that Mars must, in the past, have been very different. Or at least so some say.

When I was about to get a glimpse of Mars through the 24-inch refractor that had been used by Lowell, I suppose I was hoping in a way that I might also get a glimpse of Lowell's Mars: canals, oases, great cities, fertile corridors. Of course I did not. What I saw was a planet of light and dark, 'albedo features' as planetary scientists call them, corresponding to nothing more than vague differences in surface mineralogy and reflectivity – and most of all to the shifting seas of dust that carpet the Martian surface.

A few days before my visit to the Lowell Observatory, I had been 800 km (500 miles) away, at Caltech in Pasadena. There, I attended the Sixth International Mars Conference in July 2003. More than 400 scientists from all over the world had gathered to discuss the latest findings from telescopic observations and, most particularly, from the space probes that are currently in orbit around the planet. We heard about the latest theories – theories of water and life on Mars that have recently made the headlines.

So what was new? In the summer of 1996, when an excited NASA announced to the world that a meteorite called ALH 84001 contained what looked like fossil bacteria, the world started to go Mars-mad. Bill Clinton, no less, hailed the discovery as one of the most important in the history of science. The trouble with all this excitement was that it was based on precious little evidence. Every four months or so (as shown by some random Internet trawling), the consensus seems to switch from 'life' to 'non-life'. The ferrous crystals and tubules in the rock could be no more than wholly inorganic features.

Viking came before the meteorite; it also came before the hysteria of the 'face on Mars', although it was the cause of it. When one of the Viking orbiters took a picture of a low knoll on the Cydonia plateau, a few scientists on the imaging team joked that it looked a little like a human face. The illusion was helped by the black dots that are an artifact of the imaging process – dots that just happened to fall where the 'eyes' and, more importantly, the 'nostrils' would be. And yes, the Cydonia hill does indeed look a bit like a face, although it is more simian than human. But this was the mid-1970s, before the days of the Internet (at least the Internet as a public phenomenon), and hence the face-on-Mars story was slow to emerge. This didn't stop the inevitable conspiracy-theorists, though. But by the 1990s, after simmering away for a couple of decades, the face-on-Mars furor really took off, largely thanks to the efforts of Richard Hoagland – who was with me as I peered at Mars through Lowell's telescope. Hoagland has made a living from his books outlining the 'evidence' for ancient Martian civilizations. Martian faces are one thing, but quite another was a paper published in *Science* on 30 June 2000, in which Mike Malin and Ken Edgett, who are in charge of the camera on board the Mars Global Surveyor space probe, presented a set of images showing what they called 'gullies' on the surface of Mars. These gullies, beautifully photographed pouring forth down several crater walls, were seen as evidence that liquid water may, to this day, flow on the Martian surface.

The mystery of the gullies has dominated Martian science for the past three years. Dozens of papers have been written and presentations given, expounding the various theories of what could have caused them. Some say that water is being forced under pressure from deep underground, exiting at strata intersecting with the crater walls, rather like spring lines on

This is one of the highest-resolution images ever obtained from Mars orbit, and shows gullies in the wall of an impact crater in Sirenum Terra. It has a resolution of just 1.5 m (5 ft) per pixel. The gullies, which were discovered in 2000, have fuelled one of the greatest scientific debates about Mars since the days of the canalists. Many researchers believe they so closely resemble the gullies carved on terrestrial slopes by rain and springs that they can only have been formed by liquid water – an exciting conclusion because it suggests that Mars may be wetter, if not warmer, than has been thought for nearly half a century. Others disagree, pointing out that it is extremely hard to get liquid water to exist on the surface of Mars, and so the gullies must have been made by something else – maybe carbon dioxide gas, liquid CO_2 or water ice.

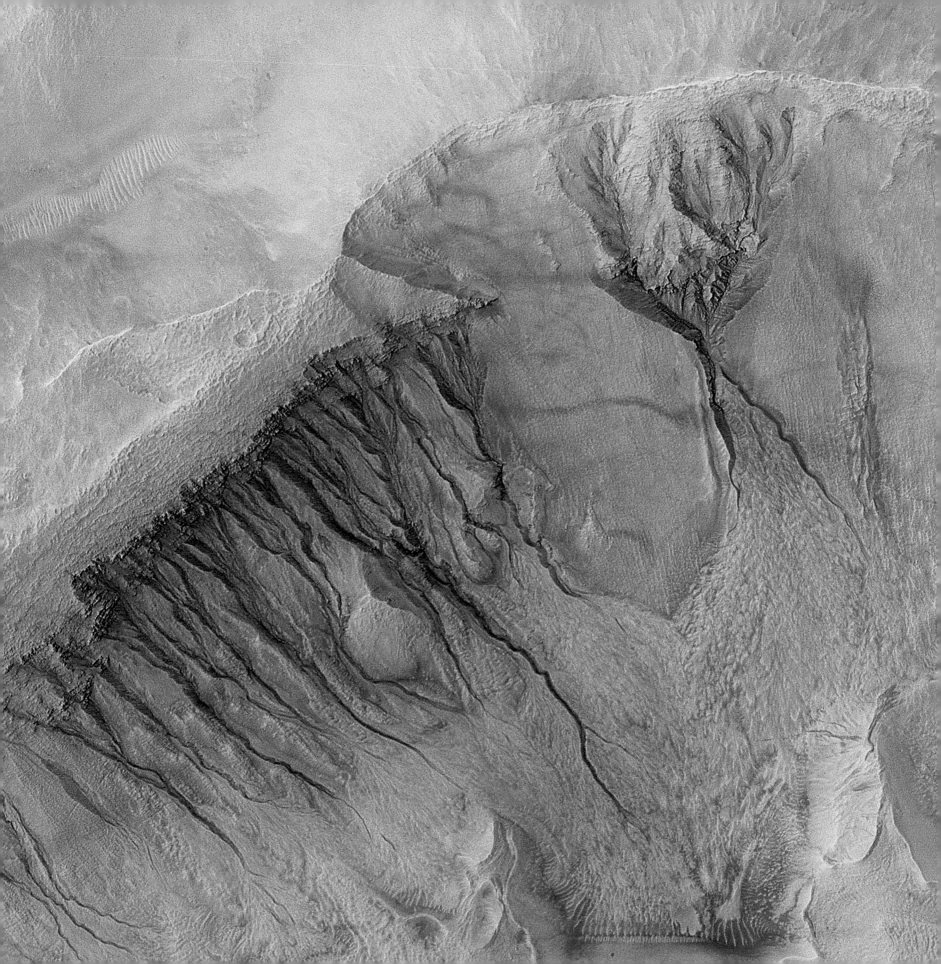

Earth. Nick Hoffman, an Australian geologist, thinks that the gullies are formed by carbon dioxide. His 'White Mars' theory dismisses the concept of large quantities of liquid water flowing on the Martian surface at any time in the planet's history, proposing instead that most of the recent features, such as the gullies, can be explained purely in terms of carbon dioxide. Then the aqueous brigade hit back. Philip Christiansen, in early 2003, claimed that melting snowpacks could have made the gullies. Vic Baker, another Arizona scientist, agrees, calling Hoffman's theories 'bizarre'. Malin, meanwhile, thinks that Mars is undergoing climate change right now – and that short-term cyclical variations in pressure and ambient temperatures may be enough to explain the existence of recent fluvial features. Others question these claims, pointing out that the mechanism for Malin's global warming – *en masse* sublimation of the ice caps – cannot work because there simply isn't enough carbon dioxide locked up in the caps, which are now known to consist mostly of water. (Sublimation is a change in state from solid directly to gas, without a liquid stage in between.) If water did flow on ancient Mars, then some serious tinkering needs to be done to the planet to make this possible. Mars would have to have been much warmer, and its atmosphere much thicker, for this to happen. Plausible mechanisms include massive volcanism melting large amounts of underground water and subsequent outpouring of gases into the Martian atmosphere. An alternative theory is that methane, one of the most powerful greenhouse gases, kept early Mars warm. As methane is not stable, it must have been constantly replenished – and this means life (in fact, recent observations suggest mysteriously large quantities of methane in the *current* Martian atmosphere). So the early Mars may have been warm and wet because there was life, not the other way round. The science of Mars is confusing and, often, contradictory.

Today, scientists talk of Mars as a world coming back from the dead. The Mariners saw a frigid, desert planet. Now there is much talk of a warmer, wetter, more benign Mars that might have existed in the distant past. Advocates of the 'warm wet Mars' theory say that the only way of explaining the ribbon-like channels, braids, and meanders that snake across the Martian surface is to invoke a world where atmospheric temperatures and pressures once allowed water to flow. Four billion years ago, during a Martian era called the Noachian, the planet must have had a very different climate to that of today. Perhaps this Eden-like Mars was home to life, say the optimists. If we send robots or humans to crack open the rocks, we will find fossils – plants, microbes, maybe even animals. New data from the Mars Odyssey spacecraft suggests vast frozen oceans of water are locked into the Martian soil. The first pictures sent back by the Opportunity rover, in January 2004, showed layered rocks that were seized upon by some as evidence that water once flowed. Maybe the planet is not geologically dead. Maybe there are islands of geothermal warmth, hot springs even. Maybe the gullies are manifestations of the flickering fires that keep Mars alive to this day. NASA has certainly bought into this idea. Its motto now is 'follow the water'. Where once the agency held back from talk of alien life, now the concept is discussed with gusto. Where there is life, there is hope.

Hope of what, though? Why do we want to find life on Mars, or at least establish that conditions there could once have allowed life to flourish? In his book *Lowell and Mars* (1976), William Graves Hoyt points out that the Martian controversy has little to do with the planet and much to do with the advocates of one position or another:

> If Lowell's popular prestige as an astronomer was largely a result of the Mars furor, the furor itself, in turn, was largely of Lowell's own making. This is to say that his long vigorous championship of his theory of life on Mars, together with the often vehement reactions this theory provoked, was primarily responsible for converting what had been a formal controversy among astronomers into a brief but worldwide cause célèbre; that it was not Mars itself, but Lowell's ideas about Mars, that caused all the fuss.

MARTIAN LANDSCAPES

Two centuries ago, the planet Mars was an orange disk with shifting greenish patches and glistening polar caps, first discovered in 1666 by Giovanni Cassini, observing from Italy. As the planet grew bigger in our eyepieces, new features came into focus. Some were real, such as the occasionally brilliant white patch known as the Clouds of Olympus (Nix

Olympica – finally discovered to be a huge volcano and renamed Olympus Mons); many other features were no more than figments of the imagination. The canals, of course, were illusory, but so too were the 'polar sea' and the lilac layer in the atmosphere that were the subjects of so much debate for most of the twentieth century, yet vanished into the sands as soon as the first space probes arrived. Even the great shifting patches of green and brown, the 'waves of darkening' once thought to be of so much import and presumed to be seasonal invasions of some sort of vegetation, turned out to be no more than the thin, drifting veneer of dust that moves around with the Martian seasons. By the end of the nineteenth century, decent maps of Mars had been produced, most notably by Frederick Kaiser in 1864, Camille Flammarion in 1876, and Giovanni Schiaparelli in the 1870s and 1880s.

Today, of course, we know an awful lot about Mars. Much of its surface has been photographed in such detail that objects the size of a family car can be seen. Landers have taken detailed measurements of its surface and have prodded, probed, drilled, and dug their way into what is becoming a comprehensive understanding of Martian geology. But there is much that remains a mystery. We may have explained the waves of darkening, but what about the mysterious dark streaks that have been seen from orbit? Are these huge splurges of salty, mineral-laden water cascading down crater walls and mountain slopes, or are they just another manifestation of Mars's shifting dust? All over Mars there are cliffs riven with gullies and 'alcoves', strange white rocks that stand out against the general ocher/red landscape. There are areas covered with craters that look as old as the Moon, and places where the surface looks so fresh that it could have been formed only yesterday. The polar caps are a swirl of water and dry ices, pitted into strange and unworldly shapes. Dunes march across the surface for hundreds of kilometers yet, unlike the great sand-seas of the Sahara or Kalahari, they appear frozen in time, locked to their rocky underpinnings. The closer we look at Mars, the more we see – but paradoxically, it seems the less we understand. The gullies could have been formed by water, carbon dioxide, or simply gravity. What is the true story of those polar pits? Why have those dunes stopped moving? Does it ever snow on Mars? We now have data beyond the wildest dreams of the early telescopic astronomers, but we also have more questions than they ever thought of in their darkest nightmares.

So what do we know about Mars? What are the incontrovertible truths about this vexing planet? A cursory glance at the landscapes of Mars shows that this is an extremely complex planet, with a long and tortuous history. It is far from being the lunar analogue where 'nothing ever happens', as it appeared to be after the first flyby, of Mariner 4. What follows is a rough guide to the planet's vital statistics – and of some of its most eye-catching scenery.

Mars is a little over half the size of the Earth, some 6,800 km (4,200 miles) across at the equator, but has almost exactly the same land area as our planet. Its density, at just under 4 grams per cubic centimeter, is eight-elevenths that of the Earth. Mars is the least dense of the terrestrial planets, and its internal structure must be quite different to those of Earth, Mercury or Venus. Mars is wholly alien and inhospitable. An unprotected human would die gasping, as the air was sucked out of their lungs and the liquid in their mouth began to boil in the near-vacuum. Until the 1960s, it was assumed that noonday equatorial temperatures regularly exceeded 20°C (68°F), a balmy summer afternoon in terrestrial temperate latitudes, but the meteorological reality is considerably more inclement. It rarely exceeds freezing point anywhere on Mars, and typical midday temperatures are usually closer to –25°C (–13°F) and plunging to –90°C (–130°F) at night. Surface pressures range from a paltry 6 millibars to an almost as paltry 12 millibars in the lowest depressions. The Martian air is almost pure carbon dioxide.

Mars is not as alien as the Moon, but it is nearly as inhospitable. No human could survive there for more than a minute or two without a spacesuit. As well as the cold, the near vacuum that is the atmosphere, and the atmosphere's toxic unbreathability, there is the matter of huge amounts of solar ultraviolet radiation raining down from above. This is not a likely hindrance to human occupation of Mars, however, as a few centimeters of soil piled over whatever habitation the astronauts have to rely on should provide an adequate shield.

What is Mars made of? Its surface, as has been gleaned from landers and orbital observatories, consists mostly of volcanic rock – basalt. Basalt is a gray-greenish rock, but Mars is red, so what we see is not unaltered, cooled lava. In essence, Mars is red because it is rusty; aluminum and iron oxides are responsible for its vivid color. Why it is rusty is a matter of much controversy. Some scientists maintain that only large-scale and long-term exposure to water can account for

the rusting of Mars. Others maintain that long-term exposure to solar radiation and the small amount of water vapor in the atmosphere could easily account for the slow oxidation of surface rocks. A new theory suggests that Mars' redness is due simply to the fact that the early Mars was too cool to allow its molten iron to sink to the core as happened on Earth. Far more iron remained in the crust and slowly oxidized. As well as basalts, parts of the Martian surface at the poles are covered with ices, both water and carbon dioxide. Deposits of olivine-rich lavas have been discovered spectroscopically from orbit, suggesting that these areas at least can never have been exposed to water for any length of time (olivine is unstable in the presence of water). Conversely, the gray hematite deposit at the Meridiani landing site chosen for the second of the 2004 NASA rovers was considered to be indicative of ancient water. Although the Martian landscape consists in large part of rugged, rock-strewn boulder plans, substantial areas are covered with fine wind-blown sand and dust. Martian dust is very fine indeed – think of something more like a liquid than even the finest talc. Experiments performed by the Viking landers suggested that this dust is chemically highly reactive, presenting perhaps a serious hazard for future astronauts.

Mars, of course, has no ocean with which to mark a datum 'sea-level'; instead, radar and visual mapping has been used to create a 'geoid' corresponding to an average surface elevation from which all surface topography can be measured.

Globally, the most striking feature on Mars is its global dichotomy. The southern highlands are on average some 3.5 km (2 miles) higher than Martian 'sea level', whereas the northern plains are about 2–3 km (1.25–1.75 miles) below it. There is therefore a 6-km (3.75-mile) difference in elevation between Mars's highest and lowest areas. Absolute altitude differences are much higher, however; the top of Olympus Mons is some 25 km (15 miles) higher than the floor of Valles Marineris. This is more dramatic than anything on Earth: the top of Mount Everest is some 19 km (12 miles) above the deepest ocean trench. The drama of the Martian relief is accentuated by the small size of the planet compared to Earth.

Interestingly, this is the one aspect of the real Mars that is far more dramatic than anyone suspected a hundred years ago, when a lack of visible protuberances at the limb of the planet was thought to be evidence that the planet was almost entirely flat. What caused the dichotomy is not yet known. It is possible that the early Mars was geologically active, with the beginnings of crustal fragmentation being seen in the Valles Marineris, which have been interpreted as a huge failed rift along the lines of the East African Rift Valley. Another possibility is that the infant Mars was subjected to a cataclysmic impact from a large body several hundred kilometers across, which partially melted the planet, which only later cooled into two distinct geological provinces.

Whatever process caused the formation of these two provinces, it must have happened very early on in Martian history, probably before 3.5 billion years ago. The southern highlands are marked – indeed defined – by heavily cratered, almost lunar terrain. It was these highlands that were glimpsed, purely randomly, by Mariner 4 during its swift flyby of Mars in July 1965. It was these highlands that – briefly – swung the compass needle of prevailing scientific opinion 180 degrees around from the wet and fecund Mars of Lowell to the late-1960s cold, dead Mars paradigm. The fact that Mars was cratered should have come as no surprise: the idea had been suggested by both Ernst Öpik and Clyde Tombaugh, who suspected that many if not all of Lowell's 'oases' may be the sites of cosmic impacts. Perhaps the most unexpected finding of the early space probes was that there appeared to be no correlation between landscape features – craters, mountains, and so on – and the albedo variations that had been the cause of so much speculation during the telescopic era. It was as if the continental outlines on the Earth, which would be clearly visible in any Martian telescope, disappeared upon closer inspection. The northern plains, as well as being lower than the southern highlands, are extremely flat, and their lack of craters suggest that the surface here has been reworked since the violence of Mars's youth, either by sedimentary processes or by lava flows. All the maps so carefully drawn by telescopic observers turned out to be illusory.

The lack of a substantial atmosphere and the extreme cold put paid to any notions of Mars being a home to life – 'proper' life at least – and NASA faced something of a funding crisis for continued Mars exploration as a result. The dis-

(Opposite) Martian mists. This Mars Orbital Camera image was taken on 13 July 2003, and shows a crater enveloped in fog. Winds from the south-east are blowing over the crater, causing the fog to ripple over its surface.

covery that Mars was less Earthlike than had been supposed led directly to a loss of interest in the planet, among the public at least.

But, as it turned out, there is far more to Mars than a simply dichotomy. For a start, north of the highland/lowland boundary there stands an enormous volcanic bulge – the Tharsis plateau – which roughly straddles the equator centered at around 115°W longitude, and which soars a lofty 11 km (7 miles) above Martian 'sea level'. No such plateau exists on Earth, and the closest comparison in extent and elevation is the Tibetan plateau, although a much closer geological analogue is the Hawaiian archipelago, which is a huge volcanic pimple on the floor of the Pacific Ocean.

Scientists do not really understand what Tharsis is. On and around the bulge stand several colossal volcanoes, Olympus Mons – which is 600 km (375 miles) across and nearly 21 km (13 miles) high – Ascraeus Mons, Pavonis

(*Above*) These maps are global false-color topographic views of Mars produced by the Mars Orbital Laser Altimeter aboard Mars Global Surveyor. The right view is of the Hellas impact basin (in purple, with a red ring of high-standing material). The other features the Tharsis rise (in red and white).

(*Opposite*) The Argyre impact basin, a vast scar in the southern hemisphere. Outflow channels run from the northern rim of the basin to the Chryse outflow region. The eastern part of the Valles Marineris can be seen at the top of the figure, near the planetary limb. The image was produced by combining data from the Mars Orbital Laser Altimeter aboard the Mars Global Surveyor spacecraft with color from a Viking image mosaic. Illumination is from the west.

Mons, Alba Patera, and Arsia Mons. There are also volcanoes around the Hellas Basin – an enormous scalloped impact crater in the southern hemisphere which sometimes fills with frost and gleams like a third polar cap – and in the Elysium region.

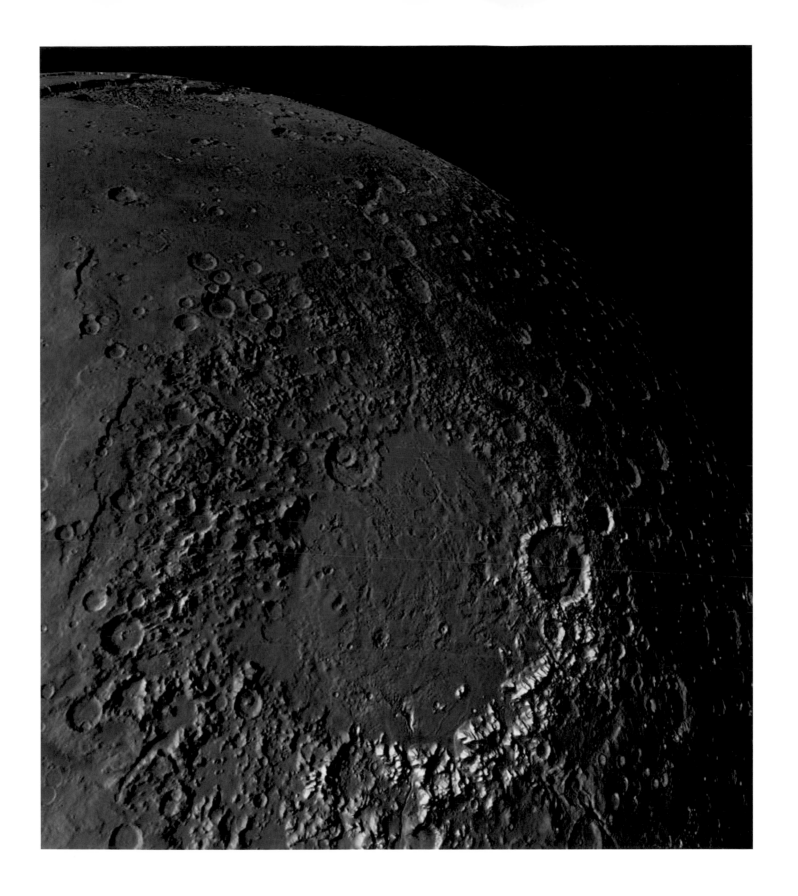

Tharsis forms a massive geological 'province', which at some 4,000 km north–south and 3,000 km across (2,500 by 1,800 miles) is akin to the area of the western United States, or most of Europe. The bulge rises steeply from the northern plains, but its southern flank is much gentler. The volcanoes of Tharsis are not active today, but they provide the clearest evidence of past Martian tectonism. The only planet in the solar system known to have active plate tectonics is the Earth. The crust of our planet is fractured into about two dozen plates, which sit atop hot, semi-molten rocks and are pushed this way and that by convection currents in the mantle. Where plates collide, the edges and the sediments which have collected upon them are bunched up into mountain ranges, where they are pulled apart, and lava wells up from below like blood into a wound, eventually solidifying and filling the gap.

Mars was too small for planet-wide plate tectonics to ever get going. Being small, more heat leaked out more quickly than from the Earth after its formation, and the Martian crust solidified more comprehensively forming a thick shield many tens of kilometers thick, too thick to be broken by heat rising from the mantle. What happened at Tharsis appears to have been a failed attempt by the mantle to punch a hole in the crust of Mars. The current hypothesis is that Tharsis was formed by a mantle plume, a rising current of magma in the slushy material below the Martian crust. Blobs of hot magma rose along this plume, striking the bottom of the cooler solid crust, forcing it upwards and in turn heating and melting its underside. It is this secondary molten rock – not the ascending mantle lava itself – that is exuded onto the surface through volcanic vents. It may be that Tharsis has been the scene of eruptions since the very early days of Martian history. Mantle plumes occur on Earth – the best-known example is the series of volcanoes that form the Hawaiian archipelago – but on Earth the crustal plates are on the move. On Mars, this does not happen, so the Tharsis plume has caused the bulge to keep on growing for maybe 4 billion years. It has been estimated that 300 million cubic kilometers of material (70 million cubic miles) was erupted in the Noachian Era alone.

Some scientists have hypothesized that in its early days, when the Tharsis volcanoes were at their most active, trillions of tonnes of gas would have been poured into the Martian atmosphere, raising the atmospheric pressure considerably and hence increasing the likelihood that large bodies of standing water could persist. This is one favored mechanism for the paradigm known as 'warm, wet Mars' – the theory that many surface features can be explained only if the climate was once much warmer and the air denser than today, and where large quantities of liquid water once existed on the Martian surface. How recently volcanoes have been active on Mars is a matter of some controversy. Detailed analysis of crater density on and around some lava flows has suggested that they were formed around 10 million years ago. Since this represents just 1/450 of Martian history, it is highly unlikely that this activity stopped just then. Many scientists believe that the volcanoes of Mars still have some life in them, and that we could even witness full-scale eruptions in the present time.

The largest volcano on Mars is Olympus Mons, discovered only by Mariner 9 in 1971. The bright spot known as Nix Olympica was a telescopic object, but its topographical reality was revealed only during this spacecraft encounter (and the mountain was renamed appropriately). Olympus Mons was revealed as it poked its mighty summit through the slowly clearing dust storm that had plagued the early weeks of the Mariner 9 encounter.

The mountain itself is a huge beast, by far the biggest volcano (or single mountain of any nature) in the solar system. Thanks to the laser altimeter aboard Mars Global Surveyor, we actually know its precise height – 21,287.4 meters (69,841 feet) – to a greater degree of precision than for any terrestrial mountain before the era of satellite mapping. This is a mountain three times the height of Mount Everest and as wide as the whole of Tibet. Its base is marked by a kilometer-high cliff. The main slopes are exceedingly gentle, around 2–4°, so climbing this monster would be no hardship – until you reached the summit caldera, when you would have to abseil down another cliff of similar height. In fact, as has been pointed out by numerous scientists, any future visitors would be unable to comprehend the majesty of Olympus Mons: it is simply too big. There is no single point on the Martian surface where a human could stand and see it as a mountain. Only at the cliff around the base of the volcano and at its summit would you have any sense that you were on a mountain the size of Texas. Most of the time, Olympus Mons would simply be swallowed by the curvature of Mars, and disappear over the horizon.

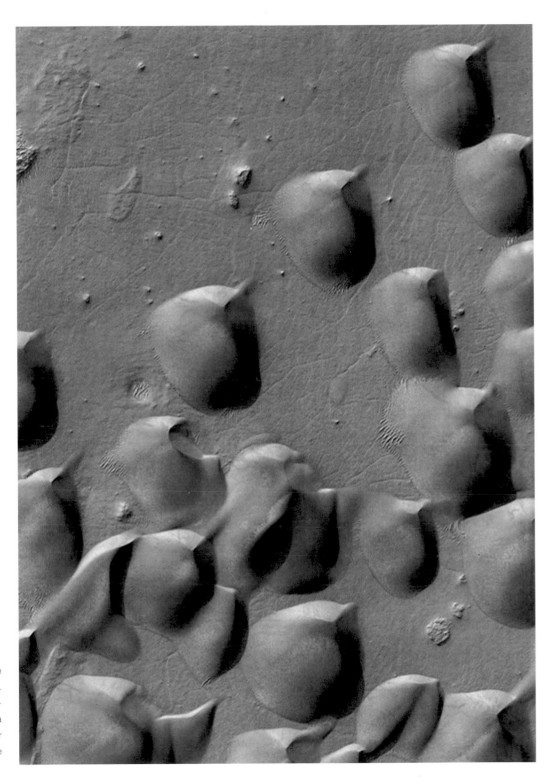

Large parts of the Martian surface are covered in huge swarms of sand dunes. These, in the crater Wirtz, were photographed by the Mars Orbital Camera on board Mars Global Surveyor in October 2002. The area covered by the picture is about 3 km (2 miles) wide.

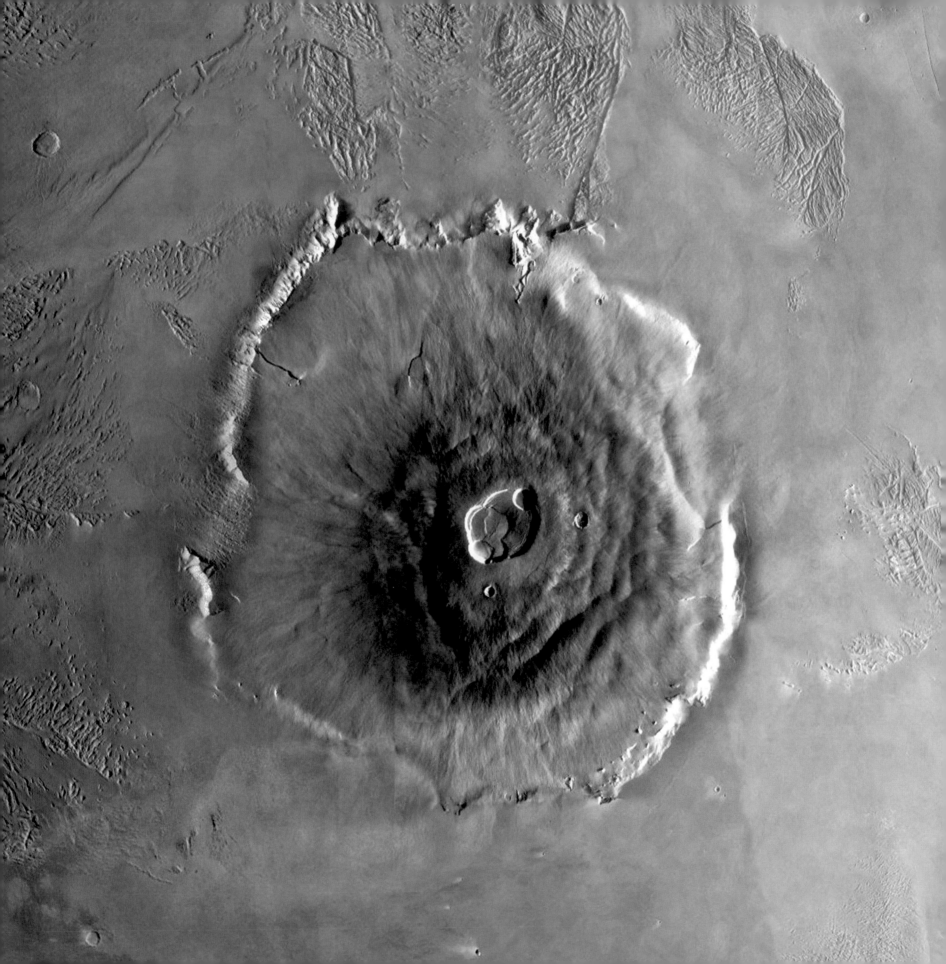

As well as the canyons and mountains, Mariner 9 and its successors have revealed that the Martian surface is home to true valleys – not canals of course, but vast, sinuous channels some hundreds or even thousands of kilometers long. These were soon interpreted by most planetary geologists to be enormous outflow channels, formed by running water in Mars's distant past. Apart from a couple on the flank of Elysium Planitia and along the eastern rim of Hellas, the huge basin in the southern hemisphere, most of the channels 'drain' from the southern highlands to the northern plains. A huge drainage system converges on the Golden Plains – Chryse Planitia – and this comprises the Ares, Tiu, Simud, Shalbatana, Maja, and Kasei channels. These valleys are typically 500 km (300 miles) long and 10–30 km (6–20 miles) wide, making them comparable in size to some of Earth's great rivers. The nature of the channels is perhaps the greatest controversy in Martian science. They look like they were formed by water, but some scientists have posited alternative explanations, suggesting that liquid or gaseous carbon dioxide could have been responsible, or even simple wind erosion exploiting lines of weaknesses along geological faults. But it is certainly the case that the channels *look* as though they were formed by water. They have tributaries, deltas, meanders, and braided sections, erosional features like teardrop islands, and even oxbow 'lakes'.

There are in fact three distinct species of channel on Mars. There are ancient (3.8–3.5 billion-year-old) branching valley networks which to most geologists look very Earthlike. According to most theories, these channels were carved out by running water which emerged from underground. The second type of channels are the so-called outburst flood channels, which appear to gush from 'collapsed terrain' and then march for hundreds or thousands of kilometers across the Martian surface into huge shallow depressions. Finally there are the gullies, which were first spotted by Michael Malin and Kenneth Edgett, the

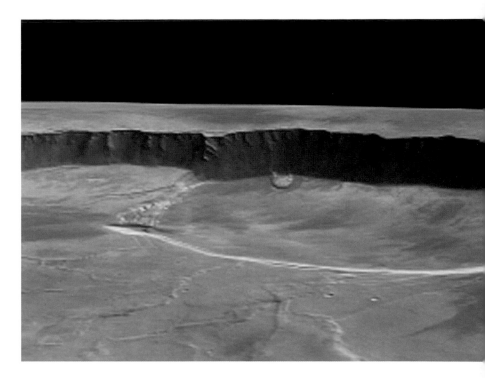

Every time a new spacecraft arrives at Mars we get a new perspective on the planet. This Mars Express image shows the power of its stereo camera, able to produce detailed three-dimensional views of some of Mars's most spectacular landscapes. Here, the mighty cliffs of the Olympus Mons caldera can be seen face on.

team running the camera aboard Mars Global Surveyor. In 2000, Malin released dramatic photographs of several 'gullies', and their nature has been the subject of fierce debate ever since.

WHAT DOES MARS LOOK LIKE?

Humans have seen the surfaces, close up, of only a handful of objects in the solar system beyond the Earth. The appearance of the Moon is familiar from the Apollo photographs and movie footage. The surface of Venus has been photographed by the Soviet Venera landers. The Venera pictures, taken under extraordinarily hostile conditions of furnace-like heat and abyssal pressure, showed a landscape of slabby gray volcanic rocks and a lowering yellow sky. To date, we have

(*Opposite*) Olympus Mons is the largest volcano in the solar system, a monster mountain almost the size of Texas. In fact, it is too big to be easily comprehended from the surface. A climber would see only a vast plain, angled upwards at just a couple of degrees when ascending the volcano's flank. Only when the giant cliffs of the caldera came into view would the true majesty of the mountain be grasped. This image was taken by the Mars Orbital Camera in April 1998. North is left, east is up.

had five 'postcards' from the surface of Mars. The Vikings, plus Pathfinder and the first Mars Exploration Rover, Spirit, showed what has become a prototypical Martian landscape scene – a rather featureless, boulder-strewn plain. These scenes resemble closely several landscapes on Earth: the Mojave Desert and the lava fields of central Iceland look superficially very similar. Despite its extremes of topography – those towering volcanoes and deep canyons – most of the Martian surface is rather flat, particularly on the northern plains, which are much flatter over a far greater extent than any terrain on Earth. Again, apart from the volcanoes, fold mountains as found on Earth are rare or non-existent. Because Mars has experienced no plate tectonics, there is none of the dramatic relief seen in fold-mountain ranges like the Alps, Himalayas, and Andes. The small hills seen in the lander photographs have mostly been the walls of craters. Paradoxically, Mars's most dramatic feature – Valles Marineris – is so grand that appreciating it from the ground would be difficult. Anyone standing on the center of the floor of the main Marineris canyon, for instance, would not be able to see the canyon walls, which are more than 200 km (120 miles) apart and therefore over the horizon. Having said that, the view from the rim of some of the 'tributary' canyons must be breathtaking, with sheer cliffs kilometers high.

The scene photographed by the second of the 2004 rovers, Opportunity, was somewhat different. The spacecraft, fortuitously as it turned out, landed slap bang in the middle of a small impact crater a few tens of meters across. The crater floor is smooth and consists of compacted, extremely fine dust, and poking out from the crater walls are white-colored rocks – possibly the first true bedrock seen on Mars. Beyond the crater rim is an almost featureless, dark-colored plane, no boulders, no dunes. The Martian sky is orange-pink, the color generated by the light scattered from fine dust particles suspended in the atmosphere. Without the dust, the near-vacuum sky of Mars would be a very deep blue, almost black.

Mars is large and varied, a difficult planet to explain. It has features that are very similar to those found on the Earth (the volcanoes, for example) and it has features that are totally alien (the pitted polar ice). There are places on Mars that could be Arizona, but there are many more that could only be, well, on Mars. Mars is not the Earth, it is true, but it is also true that, superficially at least, it is more like the Earth than anywhere else we know. Whether or not it has water and life, the two elements which NASA knows will keep public interest in the planet high, Mars is interesting in its own right. Whether we will truly understand it anytime soon is far from clear.

(*Opposite*) This three-dimensional view of the Valles Marineris, the Martian Grand Canyon, was produced by combining data from the Mars Orbital Laser Altimeter, aboard the Mars Global Surveyor spacecraft, with images taken by the Viking orbiters. The view is westward down the canyon system toward the Tharsis rise. (There is no vertical exaggeration.)

Chapter 1
Old Mars

'There is considerable but moderate atmosphere so that its inhabitants probably enjoy a situation in many respects similar to our own'

William Herschel, 1784

Percival Lowell wasn't the first to speculate on the red world. The idea that Mars was inhabited had been mooted for a long time, and observers paid close attention to the Red Planet from the dawn of the telescopic age. One of the first to learn anything useful about Mars was Aristotle. Around 350 BC, he noticed that the Moon sometimes passes in front of Mars, so Mars must be farther away. But it was not until the seventeenth century that Mars really came under serious scientific scrutiny. Galileo observed that Mars went through phases, and in 1619 Johannes Kepler used calculations of Mars's orbit to help formulate the laws of planetary motion for which he is now famous. Seventeen years later, the Italian astronomer Francisco Fontana made the earliest known drawings of Mars, noting that 'the disk of Mars is not uniform in color'. In 1666 Giovanni Cassini saw the polar caps and made the first calculation of the length of the Martian day – 24 hours and 40 minutes, just under 3 minutes off the correct figure. In 1719 Giacomo Maraldi hypothesized that the polar caps are made of water ice, and in 1784 William Herschel speculated that Mars is so apparently Earthlike – the day is almost the same length, its axis is inclined just a degree or so more than ours, it has four seasons and polar ice – that the planet must be inhabited by beings as advanced as us.

In a wonderfully incoherent essay on the denizens of the solar system (*Universal Natural History and Theory of Heaven*,

1755), Immanuel Kant speculated on men from Mars – and Venus too. It starts quite promisingly: 'In my view,' he writes, 'it is a disgrace to the nature of philosophy when we use it to maintain with a kind of flippancy free-wheeling witty displays having some apparent truth, unless we immediately explain that we are doing this only as an amusement.' So far, so good. He then goes on to explain that his readers will not be treated to any vague and unsubstantiated guesswork, only propositions 'which can really expand our understanding and which are at the same time so plausibly established that we can scarcely deny their validity.' Sadly, this initial rigor soon evaporates as, just a few paragraphs later, he tells us, 'However, most of the planets are certainly inhabited, and those that are not will be in the future.'

William Herschel also speculated on who or what may be crawling over the Martian surface, as had Bernard de Fontenelle, secretary of the French Academy of Sciences, who in 1688 wrote eloquently of a rather pleasant-sounding planet, with phosphorescent mountain ranges and strange luminous birds. In the nineteenth century, astronomers were so sure that Mars was inhabited that some even speculated on ways of communicating with the Martians, such as digging huge trenches in the Sahara Desert, filling them with gasoline or somesuch, and setting them alight. Another idea was to construct large mirrors to flash sunlight at the Martians.

Then along came Giovanni Schiaparelli, director of the Brera Observatory in Milan, who, together with Lowell, gave us the modern Mars. On 5 September 1877, he turned the observatory's expensive new 8.75-inch (220-mm) refractor on the planet, which was then at a very favorable opposition, and began to map the surface of the planet. The most prominent of the features he

'Most of the planets are certainly inhabited, and those that are not will be in the future'

Immanuel Kant

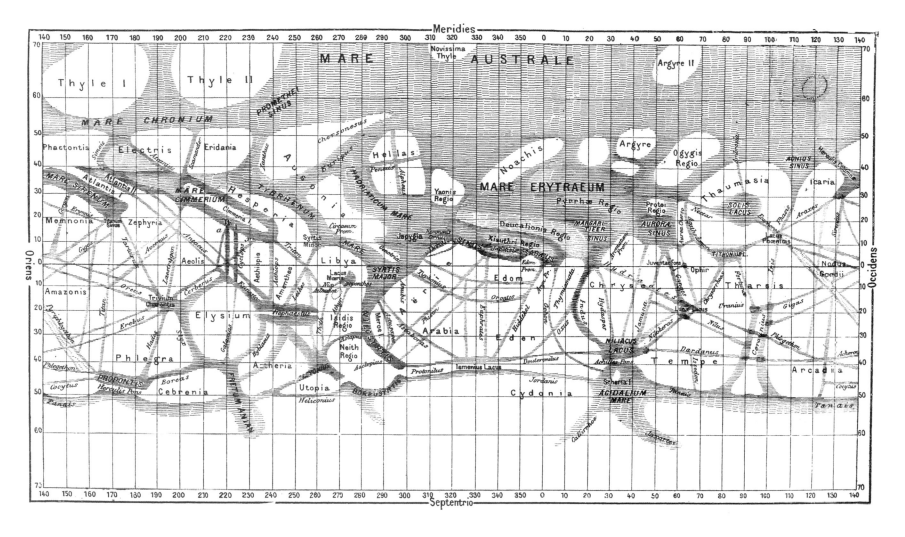

Lowell may have popularized the idea of Martian canal-builders, but it was Giovanni Schiaparelli who first saw the network of straight lines on the surface of the planet that he called *canali*. This map shows many of the features that astronomers still recognize today. The canali, however, have vanished into the sands.

saw have survived the age of space exploration. For instance, he recorded an enormous, shining oval beauty spot in Mars's southern hemisphere which he called Hellas. This we now know is the site of a huge impact crater, and sometimes it glows as brightly as the Martian poles with reflected frostlight. But what Schiaparelli became famous for was the faint network of lines that he recorded on his map. Streaking across the Martian surface, they mostly followed great circles, or gently curving paths. He found these linear features, which he named *canali* (meaning 'channels'), extremely interesting, having seen nothing of their like anywhere else in the solar system.

Few before Schiaparelli had seen any such straight lines on Mars, despite the fact that telescopes equal to or exceeding the power of his refractor had been in existence for some time. A few, including the English clergyman William Rutter Dawes in 1864, observed linear streaks, but no one had seen the vast and comprehensive system of channels reported by the Italian. Indeed, in 1877, he was the only person to see these features, despite Mars being not only in opposition but also at perihelion, and almost as close as it ever gets to Earth. But after another pair of astronomers, Joseph Perrotin and Louis Thollon, saw the canals in 1886, everyone started seeing them, as plain as day. By the late 1870s, the belief that Mars was inhabited had became widespread. But it was Percival Lowell who brought the canals – and the planet Mars itself – to the attention of the general public, after the opposition of 1892, when Mars and Earth were closer than at any time since Schiaparelli mapped the *canali*.

Before Schiaparelli, Mars was a world of shifting greens

This wonderful picture, showing frosty white water ice clouds and swirling orange dust storms above a vivid ochre landscape, was taken by the Hubble Space Telescope during the 2001 opposition.

say, by the weather or some great geological cataclysm. Schiaparelli saw *canali*, which to him were channels. Others thought them into canals, and then Schiaparelli came round to the later, 'sexier' interpretation himself.

After Kant, the idea took hold that Mars must be home to some sort of life. One reason for this was that Mars was known to possess a shell of air, just like our home world. As long ago as 1784, William Herschel had written:

> Mars is not without considerable atmosphere; for besides the permanent spots on its surface, I have often noticed occasional changes of partial bright belts; and also once a darkish one, in a pretty high latitude. And these alternations we can hardly ascribe to any other cause than the variable disposition of clouds and vapours floating in the atmosphere of the planet . . . there is a considerable but moderate atmosphere so that its inhabitants probably enjoy a situation in many respects similar to our own.

Herschel knew probably more about Mars than any other man or woman alive at the time, yet what he knew still amounted to very little. He assumed there was life there because he wanted to believe there was life there. An uninhabited world, a planet equipped with air and seas and clouds, icy poles and arid topics, whirling its way around the Sun like a cosmic *Marie Celeste*, was seen to demand more explanation than a thriving home to teeming hordes of aliens. The idea of extraterrestrials also fitted in with prevailing Christian belief; it would have seemed odd to suppose that God would not have populated every aspect of his realm with life.

But all this fantasy didn't mean that people weren't attending to the science. Key to life on Mars, as was recognized as early as the nineteenth century, was the nature and substance of its atmosphere. Estimates varied from 50 millibars – around 1/20 of that of Earth at sea level, to around 120 millibars. The general consensus was that the atmospheric pressure at the surface of Mars was somewhere near the upper end of this scale – roughly equivalent to that of Earth's air at an altitude of 17 km (10.5 miles) above sea level, or a little over twice the height of Everest. Even back then, nobody was suggesting that a human could breathe on Mars.

Early attempts at measuring the thickness of the Martian atmosphere depended on the observation of surface features at

and oranges, of ill-defined, oddly shaped areas that might be seas, forests, or even vast swarms of insects. But linear features were another matter. The presence of a large network of lines implied, to some, the work of intelligence, and although Schiaparelli was initially inclined to view his *canali* as perfectly natural phenomena, he later modified his view, accepting the argument from Lowell, mostly, that the Martian channels may indeed have been excavated by engineers rather than,

various times of the Martian day. The idea was that, as a feature neared the limb of the planet, it would begin to fade from view because it was being observed through progressively thicker quantities of atmosphere as the line of sight became more oblique to the surface (just as the Sun's brightness in the Earth's sky fades as sunset nears – as the Sun sinks, the air provides a thicker and thicker shield, reducing the Sun's intensity to the point where it can be fairly comfortably observed with the naked eye). One of the first proper estimates of the pressure of the Martian atmosphere was made by Gerard de Vaucouleurs at Le Houga Observatory, France, in 1939. He came up with a value of around 90 millibars, around a tenth of the Earth's pressure. This was near the upper end of the expected range.

Just what this atmosphere was made of was a mystery. In 1867, Jules Janssen, who founded the Meudon Observatory in the suburbs of Paris, took a spectroscope up Mount Etna and peered at Mars. His instruments told him that the Red Planet's atmosphere contained much water vapor – an encouraging sign for those who wanted to believe that the place was inhabited. His results were confirmed later by Hermann Vogel in Germany. The problem was that Janssen's instruments were lying. His technique – comparing the spectrum of Mars with that of the airless Moon – was perfectly sound, but because his equipment was crude by modern standards, his results were, with hindsight, meaningless. Later astronomers made more accurate measurements of the Martian atmosphere, with better instruments. Walter S. Adams and Theodore Dunham in 1933 found no oxygen in the atmosphere; in 1947, Gerard Kuiper detected traces of carbon dioxide.

Perhaps the oddest assessment of the conditions on Mars came in 1960, when three Americans, Clarence and Harriet Kiess and S. Karrer, concluded in a paper in *Publications of the Astronomical Society of the Pacific*, that the Martian atmosphere contained toxic nitrogen oxides, and that the polar caps were made of solid nitrogen tetroxide. Mars's red color, they thought, was caused by nitrogen peroxide. This new (but temporary) Mars was an unpleasant and inhospitable place. Away went the canals and blue polar caps, and in came an alien world where the polar glaciers, colored areas, migrating bands, clouds, blue hazes, and other phenomena were all various oxides of nitrogen in gaseous, liquid, and solid states. Kiess and his colleagues concluded, reasonably, that life as we know it could not exist in the heavy concentrations of toxic nitrogen peroxide found in the Martian atmosphere. For a while, the idea of a toxic Mars became quite fashionable (it made it into a couple of science fiction books and short stories). Happily, for those wishing Mars to be a possible home to life, spectroscopic measurements did not back up the Kiess *et al.* theory, and the idea of a Mars poisoned by noxious gases was consigned to history.

FICTIONAL MARS

Jupiter may be the biggest planet in the solar system, but when it comes to science fiction, Mars has traditionally loomed by far the largest. Before 1900, this wasn't the case. The first destination for fictional astronauts was, invariably, the Moon. The Earth's satellite is close, about 25 times the distance from London to Australia, and its surface is clearly visible. With the naked eye could be seen light and dark patches, and the ever-changing phases; with even a feeble telescope it was possible to ascertain mountains and craters, immense valleys and rocky escarpments. It was a world ripe for exploring. As early as AD 160, Lucian of Samosata, a Greco-Syrian and arguably the world's first science fiction writer, wrote a story called *The True History*, an account of a journey to the Moon. Faith in the old gods was then in decline, among both Greeks and Romans, and there was a desire for new stories to replace the old religious certainties. His yarn, which is actually a satire on the shallow storytelling which had (as he saw it) come to replace the classical, intellectual literature of the past, concerns a group of travelers who venture beyond the Straits of Gibraltar and are then thrown into the heavens by a waterspout, landing on the Moon, 'a great countrie in the aire, like to a shining island', as Elizabethan scholars translated his description 1,500 years later.

On the Moon, Lucian's travelers find themselves embroiled in a full-scale and rather futuristic interplanetary war between the King of the Moon and the King of the Sun over colonization rights to Jupiter, involving armies which boast such

'If the space probes keep redesigning our planets, what can we do but write new stories?'

Larry Niven, 1975

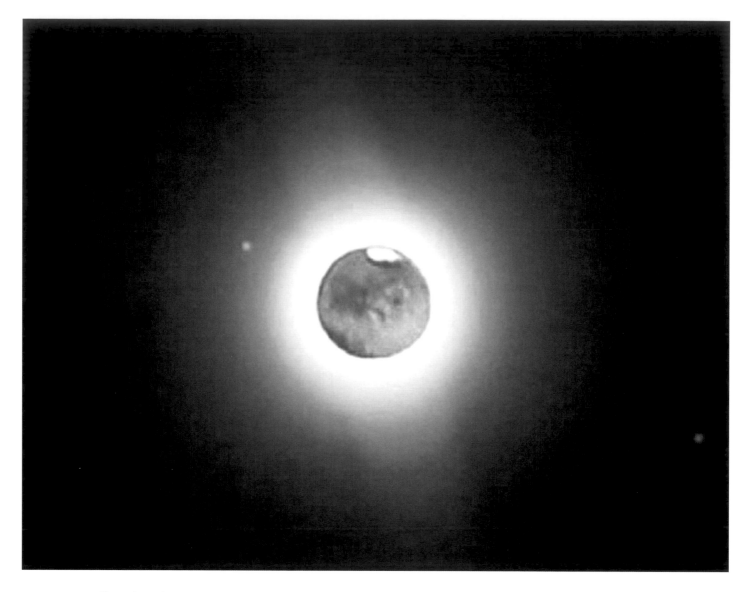

exotica as stalk-and-mushroom men, acorn-dogs, and cloud-centaurs. The human inhabitants of the Moon are also remarkable:

> Amongst them, when a man grows old he does not die, but dissolves into smoke and turns to air. They all eat the same food, which is frogs roasted on the ashes from a large fire; of these they have plenty which fly about in the air, they get together over the coals, snuff up the scent of them.

The Martian moons, Phobos and Deimos, are so small that spotting them from Earth is extremely difficult. This photograph was taken in August 2003 from an observatory in southern Austria, and shows the two moons against the planet's glare in a deliberately overexposed image – combined with a properly exposed view of Mars showing the south Polar Cap.

It is likely that few people in Lucian's time recognized that even the Moon was somewhere where one could, in theory, stand and look out into space at the Earth. In fact, for many centuries Lucian's adventure was highly regarded, not as pure fantasy but as speculative fiction, much as we might treat a science fiction novel by a respected scientist-author today. An example of this is buried in the footnotes of an 1887 edition of Lucian's work (*Cassell's National Library* series, p. 83). The original translator of this edition, Thomas Franckling, Professor of Greek at Cambridge University, writing in 1780, had this to say at the point where the Earth is seen suspended in the lunar sky as if it were itself a mere satellite: 'Modern astronomers are, I think agreed, that we are to the Moon just the same as the Moon is to us. Though Lucian's history may be false, therefore, his philosophy, we see, was true.' The editor has added, 'The moon is not habitable, 1887.'

Gulliver and the Moons of Mars – a strange coincidence.

The Laputan astronomers have likewise discovered two lesser stars, or satellites, which revolve about Mars, whereof the innermost is distant from the centre of the primary planet exactly three of his diameters, and the outermost five; the former revolves in the space of ten hours, and the latter in twenty-one and a half; so that the squares of their periodical times are very near in the same proportion with the cubes of their distance from the centre of Mars, which evidently shows them to be governed by the same law of gravitation that influences the other heavenly bodies.

This extract from *Gulliver's Travels* was written in 1726, more than 150 years before Phobos and Deimos, Mars's two tiny satellites, were actually discovered. The Martian moons, small, potato-shaped chunks of rock, are far too small to be seen from Earth except through a large telescope and under excellent observing conditions. But his details about the moons are uncannily accurate. He described Phobos's orbital period as 10 hours – in fact it is 7 hours 36 minutes, and Deimos's as 21 hours and 30 minutes – in fact it is 30 hours and 12 minutes. Not spot on, but very much in the ballpark.

Some people have delighted in reading something spooky into Swift's predictions that Mars has two tiny moons racing round it in a day or less. When, in 1877, the moons were discovered by Asaph Hall, an astronomer working at the US Naval Observatory in Washington, DC, the 'lunatic fringe' claimed that Swift must have had psychic powers, or perhaps knowledge from the denizens of Atlantis, who had developed powerful telescopes. This fringe, of course, survives to this day. One website states quite solemnly that, 'Swift must have been privy to ancient documents or knowledge. Instruments of his day could not have discerned the two moons' (www.tmgnow.com). Some cult literature has sprung up specifically to address how Swift could have 'known' about the Martian moons and has arrived at some spectacular conclusions, including the remarkable one that Swift himself was a Martian! C.P. Olivier, in his article 'Mars' written for the *Encyclopedia Americana* (1943), had this to say:

When it is noted how very close Swift came to the truth, not only in merely predicting two small moons but also the salient features of their orbits, there seems little doubt that this is the most astounding 'prophecy' of the past thousand years as to whose full authenticity there is not a shadow of doubt.

How was Swift able to predict the existence of the moons and their attributes so well? The answer is actually quite simple, and needs neither to invoke psychic powers nor to require Swift to hail from the Red Planet. In fact, he probably employed the same logic as Voltaire did a quarter century later when he also predicted two Martian moons. Voltaire knew that the inner planets, Mercury and Venus, had no moons and the two giant planets beyond Mars, Jupiter and Saturn each had many. Earth had one. It seemed likely to Voltaire that Mars, that next out from Earth, had at least two. Even if Swift had employed the same logic to figure the number of moons, his estimate of their orbital periods was still, we must admit, startlingly accurate.

After the consensus was reached that Mars was inhabited, the planet soon became the setting for numerous fantasies and romances. Percy Greg's *Across the Zodiac: The Story of a Wrecked Record* (1880) was the first science fiction novel set wholly on the Red Planet and featured a 13,000-year-old unified civilization made up of a race of Teutonic Martians. Interestingly, Greg's story was one of the few to date to take into account Mars's low gravity. In 1893, *A Parallel Unveiled:*

A Romance was published. Written by Alice Ilgenfrizt Jones and Ella Merchant, it portrayed Mars as a feminist paradise.

Wells's Mars

The most famous novel about Mars and its inhabitants is H.G. Wells's *The War of the Worlds*, published in 1898. Many assume that Wells based his writings on the near-contemporaneous observations of Percival Lowell, but that isn't so. Although Wells would undoubtedly have been aware of Schiaparelli's work, *The War of the Worlds* was written, it seems, in complete ignorance of Lowell's observations. A year after Wells's novel was published, his friend Richard Gregory, the editor of *Nature* magazine, wrote to Wells recommending that he read the works of the American astronomer. Wells's Mars is not Lowell's Mars, although there are parallels. Both are dying worlds, struggling against chronic desertification. But whereas Lowell chose to people his Mars with gentle, cooperative beings, Wells had them strike out to seek and conquer new worlds, to acquire *lebensraum* in a solar system in which the real estate had been unfairly parceled up.

The War of the Worlds really is a gripping tale. What strikes the modern reader is how contemporary it feels. The visions of a great city – London – under siege from an alien and quite terrifying power is one which has been invoked time and time again. Think of all those books, movies, and TV programs featuring a post-apocalyptic New York or London, destroyed by war or plague. In addition, Wells managed what few authors have since bothered to do –conjure up some truly *alien* Martians. His ETs were no cute humanoids, overgrown babies with funny ears or wrinkled foreheads, but terrifying beasts – malevolent and devoid of all pity and emotion. As Wells's Narrator puts it, when that first spacecraft hatch opens on Dorking Heath:

> Those who have never seen a living Martian can scarcely imagine the strange horror of its appearance. The peculiar

(*Opposite*) So vividly did Wells depict the Martian fighting machines that nearly all illustrators have come up with remarkably similar interpretations of the death ray-equipped tripods.

V-shaped mouth with its pointed upper lip, the absence of brow ridges, the absence of a chin beneath the wedgelike lower lip, the incessant quivering of this mouth, the Gorgon groups of tentacles, the tumultuous breathing of the lungs in a strange atmosphere, the evident heaviness and painfulness of movement due to the greater gravitational energy of the earth – above all, the extraordinary intensity of the immense eyes—were at once vital, intense, inhuman, crippled and monstrous. There was something fungoid in the oily brown skin, something in the clumsy deliberation of the tedious movements unspeakably nasty. Even at this first encounter, this first glimpse, I was overcome with disgust and dread.

In April 1896, two years before the publication of *The War of the Worlds*, Wells wrote a short essay, entitled 'Intelligence on Mars', on the possibility of Martian life for the London *Saturday Review*. It begins with a discussion of the publication, in the 2 August 1894 issue of *Nature*, by M. Javelle of his account of 'a luminous projection on the southern edge of the planet'. Wells notes that, 'no attempt at a reply was made; indeed, supposing our Astronomer Royal, with our best telescope, transported to Mars, a red riot of fire running athwart the whole of London would scarce be visible to him.' 'The questions remain unanswered, probably unanswerable', he adds, and continues:

> There is no doubt that Mars is very like the Earth. Its days and nights, its summers and winters differ only in their relative lengths from ours. It has land and oceans, continents and islands, mountain ranges and inland seas. Its polar regions are covered with snows, and it has an atmosphere and clouds, warm sunshine and gentle rains. The spectroscope, that subtle analyst of the most distant stars, gives us reason to believe that the chemical elements familiar to us here exist on Mars. The planet, chemically and physically, is so like the Earth that, as protoplasm, the only living material that we know, came into existence on the Earth, there is no great difficulty in supposing that it came into existence on Mars.

This passage contains several quite startling leaps of faith. How could Wells possibly have known what the weather is like on Mars? He couldn't, and didn't – he was just guessing, and guessing badly. His contemporary, Alfred Russel Wallace, surmised correctly that Martian temperatures were sub-Antarctic, not balmy. As to life, Wells assumes it exists. In this he is not so far out of tune with those scientists who today believe that life is probably universal: where it can evolve, it will.

Then Wells starts to consider Martians. In the late nineteenth century, although a plausible mechanism for evolution had been proposed, a correspondingly plausible mechanism for biogenesis was nonexistent. No one knew back then (and we remain ignorant today) how life arose on Earth. This did not stop scientists speculating on the course evolution might take on Mars. As Wells wrote:

> He would be a bold zoologist who should say that existing animals and plants would have been as they are today had the distribution of land and water in the Cretaceous age been different. Since the beginning of the chalk, all the great groups of mammals have separated from the common stock and have become moulded into men and monkeys, cats and dogs, antelopes and deer, elephants and squirrels.

Thus, he infers, it is reasonable to suppose that the Martians, if there are any, will be quite different in form from Earthly fauna:

> The creatures of Mars, with the slightest anatomical differences in their organs, might hear, and yet be deaf to what we speak; speak, and yet be dumb to us . . . no phase of anthropomorphism is more naive than the supposition of men on Mars. The place of such conception in the world of thought is with the anthropomorphic cosmogonies and religions invented by the childish conceit of primitive man.

So what *would* the Martians be like? Here, Wells falls back on the old cliché of Mars as an ancient world. He speculates that, given enough time, humans will evolve larger brains and smaller bodies, an idea explored in his essay 'The Man of the Year Million'. Eventually, evolution will produce, in a sentient organism, a tadpole-like creature with a grotesquely enlarged cranium and a pathetically shriveled body. The Martians in *The War of the Worlds* mark the final triumph of mind over body, yet of course they are still organic beings, as susceptible to infection by microorganism as the lowliest fish. It is probable

that if Wells had lived in the electronic age, his ancient Martians would have been sentient computers, machine-ghosts of their biological forbears.

According to the academic Darryl Jones, in his *Horror, A Thematic History in Fiction and Film* (2003), Wells, in *The War of the Worlds*, was engaging in

> an unimaginable statement of cultural relativism, positing a culture sufficiently technologically advanced as to render the English not only their colonial subjects, but their evolutionary inferiors and indeed their food – to be treated like livestock, or the subjects of vivisection or biological experimentation.

When the book was published, the Boer War was raging, and the first obvious cracks were appearing in the British hegemony.

The Martians fulfilled another function. Up to this point, the idea that another race of beings could in any way be superior to *Homo sapiens* was absurd, but Wells's Martians

> took as much notice of such [human] advances as we should to the lowing of a cow ... the Martian machine took no more notice for the moment of the people running this way and that than a man would of the confusion of ants in a nest which his foot had kicked – did they interpret our fire as we should the furious unanimity of onslaught in a disturbed hive of bees ... it was never a war, any more than there's war between men and ants.

This is, as Jones points out, a strong riposte to those who saw in Darwinism an apparent justification for the belief in the superiority of Man, and of English Man in particular.

The function of the Martians in *The War of the Worlds* is to give the English a taste of their own medicine, a reminder that complacency has set in among the inhabitants of the premier superpower. Up to 1898, warfare, for instance, had always been a somewhat limited exercise. It would not have seemed like that, of course, to the unfortunate combatants on the field at Waterloo, but the reality was that, compared with the slaughter to come – the industrial-scale warfare and the mass obliteration of civilians in the great wars of the twentieth century, fighting before 1914 was a relatively civilized affair.

These legless 'tripods' firing their death rays are from the 1953 Hollywood movie *War of the Worlds*, in which the Martian invasion is transplanted to 1950s California.

The Martians, with their impressive weapons of mass destruction and their indiscriminate determination to attack both civilian targets (the idea that large parts of residential London would be deliberately destroyed in warfare was genuinely shocking in the 1890s) and the military, were an amazingly prescient foretaste of the killing fields to come.

The Martians, in one view, could represent the growing power of Germany, or perhaps even America; both nations had economies that by now surpassed Great Britain's, although the existence of the Empire still gave the British an innate sense of superiority that it would take Passchendaele to blow away. According to Darryl Jones, the Martians, on one level, are actually the English, but seen through a very dark lens. They conquer and oppress, sweeping all before them with their superior technology. Where the Martians had their death-rays, the English had their ironclad battleships. (In the book the two weapons actually meet, in pitched battle on the Thames, and in what is one of the most memorable passages of science fiction warfare ever, the English man o'war briefly gets the upper hand over the Martian.) Jones speculates that if the Martians are the English, then the English are, in effect, the Irish. Indeed, *The War of the Worlds* can be read as a neat reversal of Swift's satire, *A Modest Proposal*. Instead of Irish babies being fattened up for consumption, it is Englishmen. Like Count Dracula, they have literally come to drink their blood. Wells, Jones says, also refers to the enforced economic emigration from Ireland, in the chapter 'The Exodus from London', in which the survivors of the Martian rout of the English capital are forced to gather on the Essex coast, in the hope of transportation to France or the Low Countries.

War of the Worlds *revisited: the Great Panic of 1938*

> Ladies and gentlemen, we interrupt our program of dance music to bring you a special bulletin from the intercontinental radio News. At twenty minutes before eight, central time, Professor Farrell of the Mount Jennings Observatory, Chicago, Illinois, reports observing several explosions of incandescent gas, occurring at regular intervals on the Planet Mars.

This is what listeners heard on their radios, all across the eastern United States, on the evening of 30 October 1938. A program of dance music by Ramon Raquello and the Park Plaza Orchestra, broadcast from New York, was subjected to frequent interruptions – starting with the one above. After this interlude, the music returned, only to be broken again by a 'newsflash', this time apparently emanating from the New Jersey village of Grovers Mill. There, listeners were told, a reporter named Carl Phillips was witnessing an extraordinary slight, a flaming object about 100 feet (30 m) in diameter, made of a 'yellowish-white' metal. The 'reporter' told what he was seeing:

> This end of the thing is beginning to flake off! The top is beginning to rotate like a screw. The whole thing must be hollow. I can see peering out of that black hole two luminous disks – are they eyes? It might be a face – It might be – good heavens, something's wriggling out of the shadow like a gray snake!

This was, of course, the famous broadcast of *War of the Worlds*, directed by the young Orson Welles. His Mercury Theater adaptation of the English science fiction classic was perhaps the most effective example in broadcasting history of the veracity technique. The story of the Martian invasion – relocated from Surrey to the New Jersey suburbs – was told in a series of fake news bulletins. Many people, tuning late into the show, were unaware that they were listening to a radio play.

What happened next has become the stuff of myth. The popular version of events is that millions of Americans became anxious or panic-stricken as a result of the broadcast. Those living in the immediate vicinity of the 'invasion' were the most frightened, although a degree of alarm was reported from all over the country. The play contained references to real buildings and places, even real people. Welles, then only twenty-three, had surpassed himself. It was a brilliant piece of theater, which has stood the test of time perfectly. But did it really send half of America scurrying for the hills?

The legend of the Welles broadcast is largely the creation of Princeton University psychologist Hadley Cantril, who in 1947 conducted an exhaustive study of the reactions of people who had heard the broadcast. He reported that several people actually claimed to have smelt the poison gas emanating from the Martian machines. Another said they felt the heat from the death-rays. One man even claimed to have climbed to the top of a building in Manhattan and described seeing the 'flames of battle'.

These people were not mad, or simple; instead, they seemed to have been overtaken by the sort of mass-hysteria that is extremely common in human societies, of all types and in all stages of development. The fact that a radio broadcast full of clues to its fictionality – events seemingly hours and even days apart being reported in matters of minutes – was taken so seriously by so many people is a testament to the suggestibility of the human mind.

A recent analogue to the Great Martian Panic of 1938 took place in India in the spring of 2001. There were dozens of reports of sightings of a strange creature that became known as the 'monkeyman', a crazed half-ape, half-human who took to prowling the suburbs of Delhi and other cities, raiding homes for food and even, in some cases, mauling their inhabitants. At one point, hundreds of individuals came forward with wounds they swore had been caused by this troublesome

The 1964 movie *Robinson Crusoe* on Mars depicted a harsh planet with unbreathable air. In reality, Mars is harsher still, and is unlikely to be explored by a lone astronaut.

simian. Of course, there was no monkeyman. A study of the phenomenon later in the year concluded that the whole episode was the result of a peculiar form of mass-hysteria.

According to Robert Bartholomew, an Australian psychologist, the legend of the Welles broadcast has become as much of a myth as that of the original invasion itself. 'The irony here', he wrote in an article published in the *Skeptical Inquirer* in 1998, 'is that for the better part of the past sixty years many people may have been misled by the media to believe that the panic was far more extensive and intense than it apparently was.' However, he accepts that, regardless of the extent of the panic, there is little doubt that many Americans were genuinely frightened, and some did try to flee the Martian gas attacks and heat-rays especially in New York.

Myth one in the Mars broadcast was the fictional invasion. Myth two, in both cases, is in many ways more interesting – that there was a nationwide panic over the invasion. The modern equivalent was the Millennium Bug; both the effects of the bug and people's reactions to it were massively over-

reported in the media. There was no doomsday, nor were there millions heading for the hills with generators – just a few gullible crazies. The story of the *War of the Worlds* broadcast has now become part of the Great Mythology of Mars, adding to the canals and the face.

Mars Attacks! Rarely have invading Martians been so silly – or so sinister – as in the wonderful and much underrated 1996 Tim Burton movie.

After War of the Worlds

Wells made Mars popular. Countless books have been set on the planet or have featured Martians in some form or other. But after Wells, surprisingly few authors bothered much with their Martians, making them more or less human with strange eyes, strange skin, or funny ears. There were some notable exceptions. Ray Bradbury's Martians (in *The Martian Chronicles*, 1951) are ethereal creatures, appearing only occasionally. Tackling racism, pollution, and Cold War paranoia, *The Martian Chronicles* was one of the first novels to fail to describe Mars through a lens of wit and often bitter irony. One theme was nostalgia – nostalgia for a Mars that never existed. When Bradbury wrote the *Chronicles*, scientists knew that Mars was home to no civilization, that it had neither crystal cities nor humanoid inhabitants. Mars was already

dying, temperatures plummeting, the atmosphere thinning out; Lowell's Mars of the early twentieth century was but a distant memory.

In the early 1950s, Mars could still be written about, albeit with a certain amount of tongue in cheek, as a warm, wet, and well-aired world where a human could live happily without the twin inconveniences of asphyxiation and freeze-dry desiccation. Valentine Michael Smith was the only survivor of a mission to Mars in Robert Heinlein's 1960s classic *Stranger in a Strange Land*, hailed as a 'mind-bending' masterpiece by the god of wry science fiction, Kurt Vonnegut. In Vonnegut's *The Sirens of Titan*, Mars has a walk-on part, home to a bizarre army of lobotomized humans, run by a mad tycoon for his own evil purposes.

When Mars was gasping for breath, after the arrival of the Mariners, science fiction writers needed to remake the planet in the new scientific image. One of the most vivid and coherent accounts of a real Mars is in Frederik Pohl's classic *Man Plus* (1976). By now, authors had started to move away from any sort of realistic depiction of spaceflight. Apollo had been and gone, and Von Braun-esque romances of vast rocket fleets were passé. Although the Vikings had yet to arrive at Mars when Pohl wrote this book, his Mars is distinctly Marinerian rather than Lowellian – cold, dry, and airless, with no canals but with plenty of sand and endless boulder-strewn plains. His solution to the problem of how to colonize Mars is both realistic and plausible: rather than grandiose schemes to terraform the place, or building huge domed cities under which humans can shelter, Pohl's space program chiefs decided that rather than remake Mars in Earth's image, it would be quicker, easier, and far cheaper to remake humans as Martians. Thus Roger Torraway, the astronaut hero, is subjected to a painful remodeling, his skin flayed off and replaced with a thick leathery hide. He is over-muscled, skilled-up, and de-sexed. Mars lurks in the background, a rather sinister, silent, and unpleasant place – as in reality it is.

Mars has not really been given a decent movie role since. *Total Recall* (1990), starring the current governor of California, Arnold Schwarzenegger, had a fair stab at representing a scientifically literate Mars. There is no air, and the colors are right, but like every single movie ever made about the Red Planet its makers didn't bother with the one-third-and-a-bit gravity. This film features trains, which are something of a staple in fictions set on Mars. The planet so much resembles the Old West that it is crying our for the Santa Fe Railroad. Fictional Mars has mirrored Scientific Mars for a century now, usually lagging about a decade behind. Films made in the future will need to tackle the great debate over Martian water because that, overwhelmingly, is what dominates all discussions about the planet today.

Chapter 2
A Question of Water

'There can be little doubt that water has played a major role in the history of Mars, but uncertainties remain about almost every aspect of the water story'

Michael Carr (*Water on Mars*, 1996)

Until 2000, Mars was cold, dry, and utterly inhospitable. Viking had failed to find life on the surface, and despite intriguing evidence from orbital cameras that water had once flowed there in the distant past, most scientists believed that Mars was locked in an almost eternal ice age, a 3.5-billion-year-plus epoch stretching from a time when bacteria ruled the Earth to the present. And Mars appeared to be geologically dead. With a surface pressure averaging 1/130 of the Earth's and temperatures of –90°C (–130°F) and below, the idea that liquid water could flow on the surface seemed preposterous. Mars was too cold and its atmosphere too rarified to allow that to happen. Mars, in short, was looking rather dull.

Then came the stunning announcement on 30 June 2000, with a paper in *Science* by Mike Malin and Ken Edgett, who run the camera aboard the Mars Global Surveyor spacecraft. The front cover of the journal showed a stunning orbital shot of two gullies on the upper rim of a crater in the Gorgonum Chaos region a site, in the Martian southern hemisphere thought to have formed as a vast quantity of water erupted from the ground causing it to slump. The news created quite a stir. According to a *Science* editorial by Richard Kerr,

it began as a whisper on the Web a week ago on Monday

(*Opposite page*) The cover image of the June 2000 issue of *Science* magazine. These 'gullies', in the wall of a crater in the Gorgonum region, were spotted by Michael Malin and Ken Edgett, who are in charge of the camera aboard Mars Global Surveyor. Many geologists interpret the gullies as evidence that water flows across the Martian surface today. Others disagree. The boulders scattered along the gullies in this image are about the size of a large family car.

evening, grew to a noisy torrent of media babble by Wednesday, and on Thursday morning crashed onto the front pages. Moving at the light-speed pace of modern media, a wave of chatter about water and possible life on Mars swept a paper at *Science* into headline news a week before its scheduled publication.

INTERESTING PLANETS V. BORING PLANETS

The most interesting planet in the solar system is the Earth. It has everything a planetary scientist could dream of: big shiny oceans, continents, plate tectonics, volcanism, and, most importantly, life. It could even be the only planet in the solar system to have been studied in detail by aliens. If such beings exist within 100 light years or so, and they have the where-withal to build large space telescopes, Earth's clear oxygen-and-methane signature will ring out like a bell.

Earth is unique in having a gigantic biosphere, but it is not unique in being interesting. The giant planets – Jupiter, Saturn, Uranus, and Neptune – are interesting because they are huge agglomerations of rock and gas with thick, active atmospheres. Some of their moons are interesting too. Titan, Saturn's largest, is planet-sized and has an atmosphere even denser than the Earth's. Jupiter's big moons are all fascinating worlds in their own right. Io is pockmarked with volcanoes – more than 300 of them at last count – and is the most geologically active body in the solar system. Its surface is drenched in radiation, and features lakes of brimstone and kilometer-high fire fountains. Europa looks superficially dull,

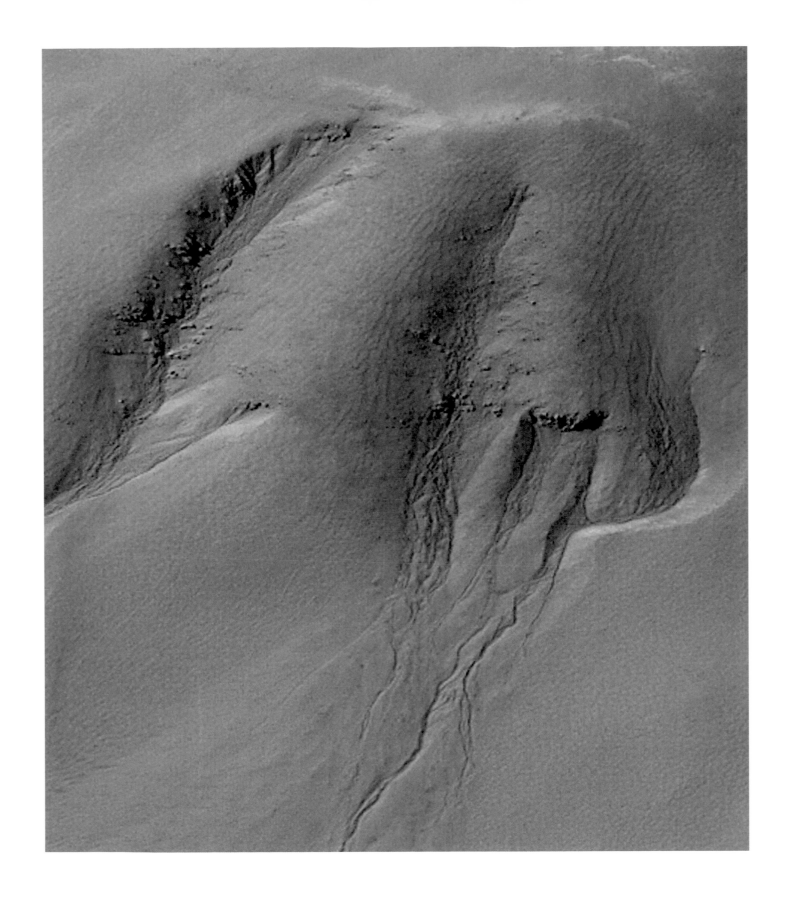

a vast ball of ice, but an estimated 20–40 km (12–24 miles) beneath its frigid surface there appears to lurk an ocean, the solar system's biggest. In fact, Europa vies with Mars for the title of most interesting object in the solar system. The possibility of life in its murky depths is taken so seriously that in September 2003 the highly successful Galileo probe, which had spent much of its operational life uncovering the secrets of Europa, was deliberately flown into Jupiter to avoid the risk of a wayward crash into Europa, decades hence, which might contaminate an alien ecosystem.

Ganymede and Callisto, Jupiter's other big moons, are fairly interesting. Ganymede has large, strange structures on its surface that cannot easily be explained, while Calliso has a magnetic signature that suggests another possible subsurface ocean, although this one is likely to be even less accessible than Europa's. Neptune's moon Triton has huge geysers, spotted by Voyager 2 in August 1989, spewing nitrogen into space. No one thought anything solid this far out from the Sun would be geologically active.

Where else is interesting? Well, Pluto may have hidden surprises – there seems to be something odd going on in its atmosphere, although we are unlikely to find out what exactly for many, many years. NASA has made repeated efforts to get a Pluto mission off the ground but so far without success

Shining like a billiard cue ball (and almost as smooth), Europa, the second of Jupiter's planet-sized satellites, is one of the most interesting objects in the solar system, rivaling Mars as a possible home for life. Beneath its icy surface is thought to lie a vast, dark salty ocean of liquid water, protected from the intense cold and radiation of the Jovian system by the ice crust above.

(a new mission to Pluto is scheduled for launch in 2006). Titan, Saturn's giant moon, will be visited by the Cassini-Huygens mission in early 2005. Scientists believe that beneath its clouds lies one of the most bizarre landscapes in the Solar System. The asteroids are interesting, not because they have oceans, life, or volcanoes, which they do not, but because of the unfortunate fact that every now and then one of their number crashes into the Earth, with regretable results that could one day include the end of our species. The comets are interesting too – as well as providing the occasional spectacular display in our night skies, they also seem to be laced with organic chemicals such as hydrocarbons, and even the precursors of life.

So what about the places that are boring? The Moon is commonly held to be boring. No atmosphere, no water (no liquid water anyway), no volcanoes, no life. Venus is boring too, but for different reasons. Its surface is a 450°C (850°F) high-pressure hell; it is unlikely that any human will ever stand

Venus, seen here stripped bare of her clouds by Magellan's radar, is one of the most inhospitable places in the solar system. With surface temperatures of 460°C (860°F) and pressures of 90 bars, Venus has been written off as a possible abode of life. Perhaps because of this it has been ignored in recent years in terms of space exploration, although ESA will launch a new probe to our sister planet, Venus Express, in November 2005.

there, although nothing can be ruled out. Mercury is also deemed dull: it looks like the Moon and, like our satellite, is pockmarked with craters. Craters are the hallmark of a boring planet. When Mariner 4 saw craters on Mars in 1967, everyone groaned. Craters equate with dullness because craters equate with age. In the early days of the Solar System, there was a lot more debris flying around than there is today. All the planets (and their moons) were hit on a frequent basis, causing massive cratering and scarring of their surfaces. Slowly, the rate of cratering declined as the supply of debris began to dwindle, and for more than 3 billion years large impacts have been rare events. Any planetary surface that is rich in craters must therefore be incredibly old; Earth was hit as frequently (in fact more frequently) than the Moon, its relative lack of

craters is testament to the erosional power of wind and water – and also the power of volcanism and tectonics to remodel the surface.

Of course, this is all nonsense. There is no such thing as a boring planet – or moon or asteroid or comet. A boring planet would be a planet about which we knew all that there is to know, and this is magnificently not the case for any object in the solar system. Take the Moon, for instance. The Moon is a mysterious place. Water ice may lurk near its poles – a radar signature of water was found by the Clementine probe in 1994, and verified by a neutron spectrometer carried aboard Prospector in 1998. There are signs that the Moon may not be as geologically dead as was thought. Everything about the Moon, from its bizarre history to the makeup of its rocks and its internal structure, is wholly and utterly fascinating. Ending the Apollo program was a scientific crime of the first magnitude.

Neither is Venus boring. It may be too hot for life, but the story of its surface is one of the great puzzles of planetary science. The Venusian crust has a fair smattering of craters, but using their density to calculate its age gives a figure of some half a billion years, which is adolescent by planetary standards. Furthermore, almost all of Venus's surface appears to be the same age. This suggests something extraordinary: that the entire surface of Venus was remodeled, in one fell swoop, 500 million years ago. This could have been the result of a combination of a thick crust, no plate tectonics, and a lot of internal heat. On Earth, heat from the interior is released steadily through the mechanism of plate tectonics. On Venus, it is possible that this heat simply builds up like a huge pressure cooker which periodically 'boils over' or even explodes.

You won't meet a single planetary scientist who will dismiss any object as boring. Yet a subtle PR game has been played with the planets since the dawn of the space age. With no possibility of life, and seemingly geologically dead, the Moon became a hard 'product' to sell to the taxpayer after the first landings. Venus too has slipped off the radar. But Mars, the classic 'alien world', has always sold itself with ease. And now, for NASA and other space agencies, an active and possibly even living world is quite literally a gift from the heavens.

An excited NASA drew the public's attention to the discovery made by Mars Global Surveyor at a press conference in the Jet Propulsion Laboratory (JPL) auditorium in Pasadena.

Taking the stand were Malin and Edgett. Malin showed his photographs and stated, bluntly, that 'had this been on Earth, there would be no question that water was associated with it'. Edgett added that the debris scattered around the base of the gullies looked fresh enough to have formed only yesterday. Cautiously, he said that it may be 1 or 2 million years old. The implication was clear: Mars, as recently as a couple of million years ago, was not geologically dead. And if it was not dead a couple of million years ago, then it was reasonable to assume that it was not dead now. After all, the chances of humankind starting to study a planet just 2 million years – less than one-twentieth of one percent of its age – after all activity had ceased were very small indeed. If these gullies were 2 million years old, there were almost certainly others in the process of being formed.

More importantly, the 'W' word had been used: water, blue gold, the key to life, excitement and fun on any planet that possesses it. To the casual observer, the gullies look like the channels that are carved in sand by Earthly streams. You can recreate Martian gullies on any beach with a bucket of water quite easily. And to the geologist (well, some geologists) they look pretty watery as well. Perhaps liquid water is, to this day, gushing out of those crater walls, carrying sand and rocks – some the size of cars – downslope with it. If there is water gushing out of crater rims, then Mars suddenly becomes a very different place, a wet world. And where there is wetness, life can exist. This was exciting news.

CAN LIQUID WATER EXIST ON MARS TODAY?

After the true thickness of the Martian atmosphere was revealed in the 1960s, it became clear that even if there were Martians and even if they wanted to build some canals, they would have a hard time doing so. The reason is that atmosphere. The Martian air is simply too thin to allow liquid water to exist. At a pressure of just 1/130 or so of Earth's, water is unstable. Most accounts of what happens to water exposed to the Martian environment state that it immediately flashes into steam, boiling vigorously before freezing. This would seem to preclude rivers, lakes,

or even streamlets carving away at crater walls to form gullies and alluvial fans. In fact, the story is a little more complicated than that. In the case of the large outflow channels, so much water may have erupted so rapidly that the rate of loss to the atmosphere was so slow that there was a sufficient quantity to erode the surface – after all, lava (hardly a stable liquid) can and does erode the surface of the Earth. It is true that liquid water is not stable on the Martian surface, but it is also true that it is not stable on much of the Earth's surface either. To understand what happens to putative Martian water, it is necessary to delve into the strange world of the physics of water, a rather strange substance.

Water boils when you heat it. That is, when a certain temperature is reached, it starts to turn into a gas. On Earth, at sea level, this happens at 100°C (212°F). But if you climb a mountain, you will find that the boiling point drops as the atmospheric pressure falls. The lower pressure makes it easier for the water molecules to escape from the surface of the water. On top of Mount Everest, for example, the boiling point of water is about 80°C (176°F), whereas at the bottom of Death Valley, which is 400 m (1,300 ft) or so below sea level, it stands at 105°C (221°F). It is impossible to make a decent cup of tea at the first location, whereas it is possible to produce a most excellent brew at the second.

On Mars, the pressures are much, much lower than even at the summit of Everest, and tea-making is not really possible at all. In fact, on Earth you would need to go to an altitude of some 26,000 m (85,000 ft) to mimic Martian surface pressures. At or below the mean datum level (the average surface elevation, which is used as an alternative for sea level in the absence of seas), because of the low surface pressure the boiling point of water is reduced to only a few degrees above the temperature at which it will freeze. The highest pressures are found at the bottoms of the Marineris canyon system and on the floor of the Hellas basin. At higher altitudes, the pressure drops below the so-called triple-point of water – that is, there is no temperature above freezing at which liquid water will not boil. Below 60 millibars, water ice passes directly to water vapor when it is warmed (a process known as sublimation). Water is a peculiar substance in that it freezes at a higher temperature as the pressure decreases – at Martian pressures this is enough to raise the melting point to a couple of degrees or so above zero celsius.

So how would water behave on Mars? At low altitudes, cold liquid water at a temperature of a couple of degrees or so will not suddenly start to boil. It will, however, start to evaporate rather fast, 'as quickly as your coffee on the warmer if it is at 60°C [140°F]' says NASA's Michael Hecht. As it evaporates, a new process kicks in: it starts to cool, perhaps at the same rate as it would cool on Earth if the air temperature were − 60°C (−76°F) or so.

In fact, the thin Martian atmosphere actually makes finding liquid water more rather than less likely, given the low temperatures. If the air were as thick as on Earth, but the temperatures were still Martian, we could be sure that liquid H_2O would be nonexistent on the Martian surface. But because the air is so thin, any liquid water will lose heat extremely slowly by convection. Water on Mars would also have some advantages over water in cold air on Earth: the evaporation rate can be reduced significantly if the air were already saturated with water (as frequently happens), if the water were salty or otherwise contaminated, or if it were covered with snow and ice. So the issue of liquid water on Mars is somewhat counterintuitive.

What all this means is that there are plenty of places where liquid water could, in theory, exist on Mars for substantial periods of time. In lowland areas, a bucket of water would develop a thin ice crust fairly quickly, but it would be many hours before all the water froze solid. At higher altitudes, liquid water is indeed unstable, but in practice there are many ways in which it could survive – such as beneath a surface crust of ice or dust, or if it contains significant quantities of dissolved minerals such as salts (which have the effect of lowering the freezing point). Liquid water can certainly exist on Mars – just as hot coffee can exist on Earth, even though neither may technically be stable. So why do so many sources state that water disappears into a puff of steam and ice the moment it is exposed to Martian conditions? Michael Hecht explains:

The dark material in this valley is sand and dust, blown into it over countless millennia. The unnamed valley is in Newton crater, and its form is certainly evocative of a water channel. The image was taken by the Mars Orbital Camera on board Mars Global Surveyor in July 2003.

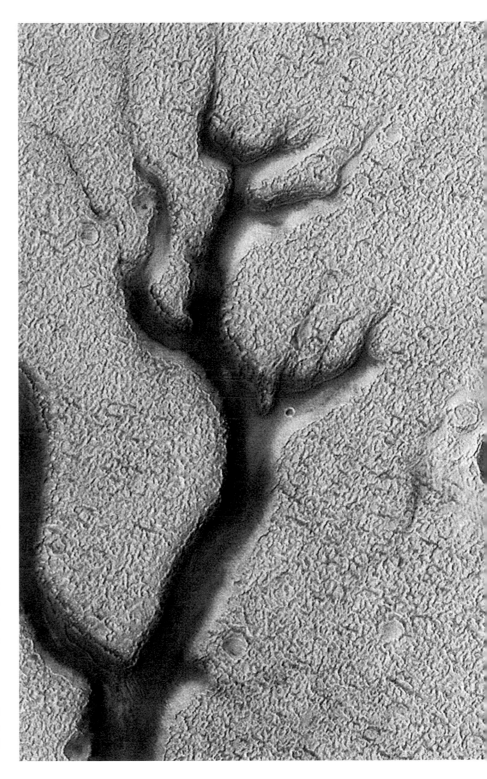

The source of this story may well be the experience many of us have had with Mars simulation chambers, such as the enormous facility at the Ames Research Center. It is common to place a container of water in those chambers as a quick visual indicator that the pressure is going down. The water does, indeed, boil violently and then freeze. The reason, however, is that the water is at room temperature, well above the boiling point at Martian pressures. So liquid water can exist on Mars. But this does not necessarily mean that it does.

Mars's pitifully thin atmosphere means that its weather is usually uneventful, but there is visual evidence of day-to-day changes. Overnight frosts were seen by both Viking landers. This image shows a thin layer of frost on the surface at Utopia Planitia. It was taken by the Viking 2 Lander on 18 May 1979. The layer is only a couple of thousandths of a centimeter thick, and is composed of a mixture of water and CO_2 ices.

THE WARM, WET HYPOTHESIS

So the question of liquid water on the surface of Mars is somewhat vexed. It seems that the laws of physics do not prevent water from existing there, but neither do they promote it. Yet there is evidence that the surface of Mars was once covered in immense quantities of water. Seen from space, the Martian deserts are bisected by huge sinuous channels, some of which are thousands of kilometers long. These channels – undoubtedly dry now – look as though they were carved by water. They even have meanders and levees. There are what look like ancient shorelines, and 'fossil' deltas. For many scientists, there is a simple conclusion: that while Mars might now be a cold, desiccated hell, it almost certainly was not so in the distant past. But the evidence is contradictory. Some of the volcanic rocks on the Martian surface look fresh, pristine. If Mars was once covered by rivers, lakes, even oceans, then geologists would expect these rocks to show the unmistakable signs of weathering – chemical attack. In fact, there are large quantities of a mineral called olivine that have been detected on the Martian surface that could not be there if they were under water for any length of time. Water – if that is what carved the channels – left its mark visibly but not, apparently, chemically. It is all very puzzling, and has led some scientists to doubt the existence of a watery past. Others point out that it may be possible to make Mars warm and wet – or at least cool and damp – for short periods, possibly following an intense period of volcanism or a meteoritic impact. Results from the robot geologists orbiting Mars suggest that Mars has always been dry. But the photographs of 'watery' features suggest otherwise, as do the findings from the latest landers. There is as yet little consensus. The question of Martian water – what role it played in the planet's history and what role it is playing today (and will play in the future) – now totally dominates the Mars question.

(*Opposite*) This Viking image shows a large dendritic (treelike, branching) valley in the southern highlands. Many argue that it is almost impossible to explain features like this without recourse to water erosion.

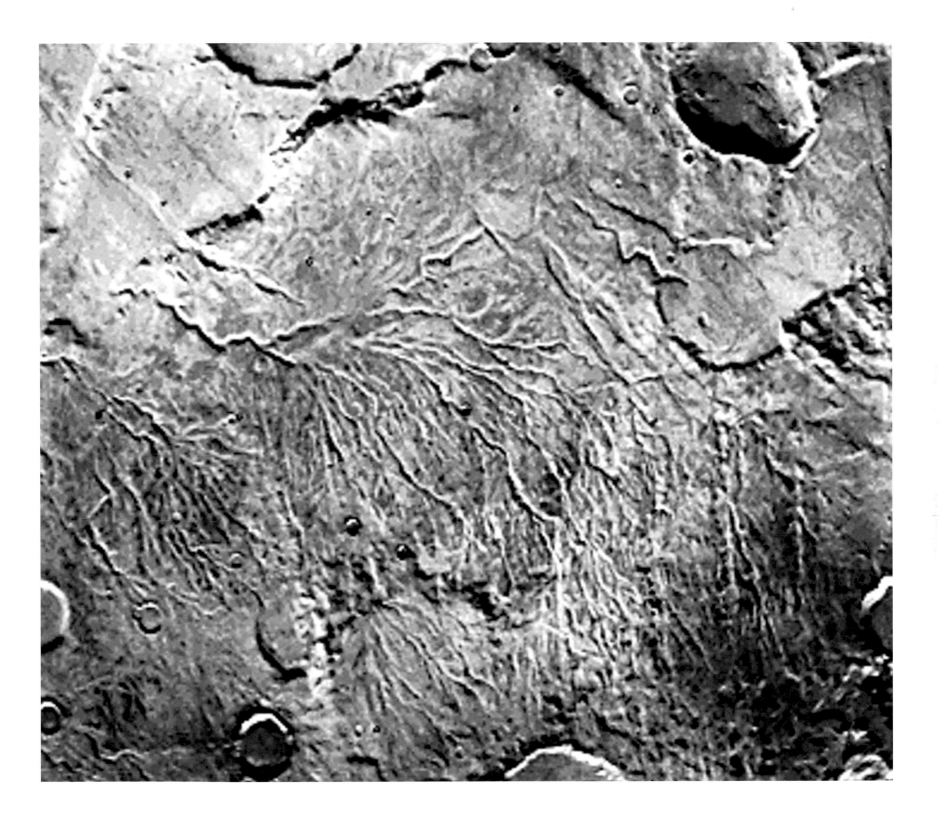

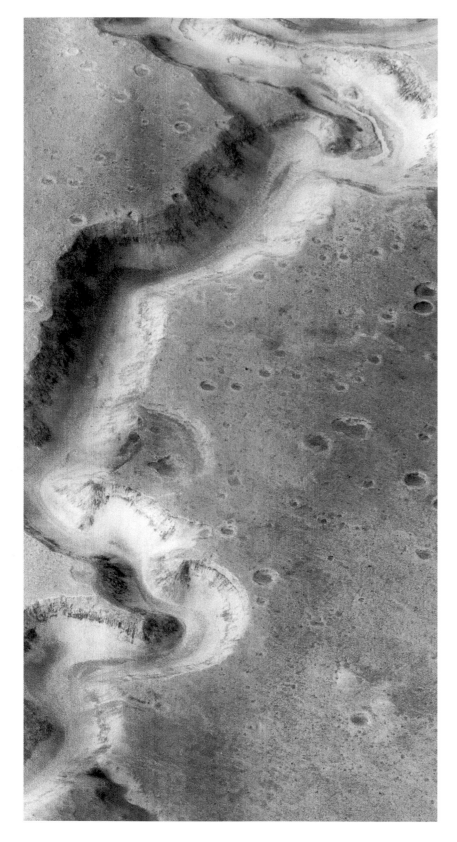

MARTIAN GEOLOGICAL HISTORY

No human geologist has ever hammered open a Martian rock. All our geological data from Mars comes from the handful of machines that have landed on its surface and the probes that have surveyed the planet from orbit. Excellent though the robots have been, there is still much we don't know. But we do know that some parts of the Martian surface are older than others. Much of the southern hemisphere is heavily cratered, implying great antiquity. The northern hemisphere, by contrast, is less so, implying that much of what we see there is younger. In fact, the division of the Martian surface into two hemispheres is extremely clear. Some places in the north, for example the slopes of some of the great volcanoes, are barely cratered at all, implying extreme youth. There are features on Mars that appear to have been created very recently, in geological terms, such as those gullies found by Malin and Edgett. The polar ice caps wax and wane seasonally. The explanation for the relative youthfulness of the Martian northern hemisphere is not clear. With no plate tectonics to recycle crustal material, the smooth plains north of the equator could be covered either by sediment or by lava.

Geologists have constructed a calendar of the great epochs of Earth history. These eras and epochs are marked by the fossils they contain and by the prevailing climates at the time. The Carboniferous was (crudely) a time of humid tropicality, of fetid swamps and deep impenetrable forests. The Jurassic and Cretaceous were the eras of giant reptiles. Immense deserts and dustbowls characterized the Permian, when the Earth's continents came together and held hands in one of their periodic coagulations to form the giant landmass known as Pangaea. Overshadowing these familiar times is the great, uncharted sea of the Precambrian, more than nine-tenths of our planet's history, dismissed in even quite recent geological textbooks as simply 'the time before'. Can we do the same thing for Mars? On Earth, a geological map is compiled from a wealth of detailed observations made at the surface, whereas for mapping other planetary bodies we have almost no 'ground truth' at all to draw upon. Can we, without a single geological field trip or even a proper aerial survey, construct a map of the Martian past?

The simplest way of describing how Mars has changed over time would be to describe the age of each part of that surface. But how are we to determine it? We cannot go down and work out the sequence of rock strata as we can here on Earth – although there seem to be some spectacular formations and exposures awaiting us. We have no way of telling, even if we do see layers from orbit, which way up these layers are. On Earth we have immediate access to the surface, and often to many other older surfaces buried beneath it. (Usually, if one surface is on top of another, the deeper surface is older – but not always. Tectonic forces may flip over the strata, so that the oldest layers are on the top. Working all this out is what we have field geologists for.) In the Grand Canyon, for example, one can see on the canyon walls the various layers of rock that have been excavated over time. The process of dating layers by looking at the relationships between them is called stratigraphy.

In addition, by using radiometric dating techniques, we can discover when a rock was formed or changed its state, by measuring the amount of materials produced by natural radioactive decay within the rock. We no longer need to work on the dead-reckoning of calculating erosional or depositional rates to work out the age of rocks. So on Earth we can see the sequence and we can date it. This powerful technique has allowed us to determine the absolute ages of all of the surfaces on Earth and those surfaces of the Moon from which samples have been returned by the Apollo and Soviet Luna missions.

After the Mariners and the Vikings, geologists thought that the rocks on the Martian surface were almost entirely volcanic. But with the arrival of Mars Global Surveyor and its high-resolution camera, hundreds of layered formations have been revealed. The interior of this ancient impact crater, in the Schiaparelli basin, about 2.3 km (1.4 miles) wide, shows many layers of sedimentary rock. These layers may have been deposited in a lake, or perhaps they were blown in by winds. It is unclear whether we are looking at actual rock here, or at layers of loose, semi-compacted dust.

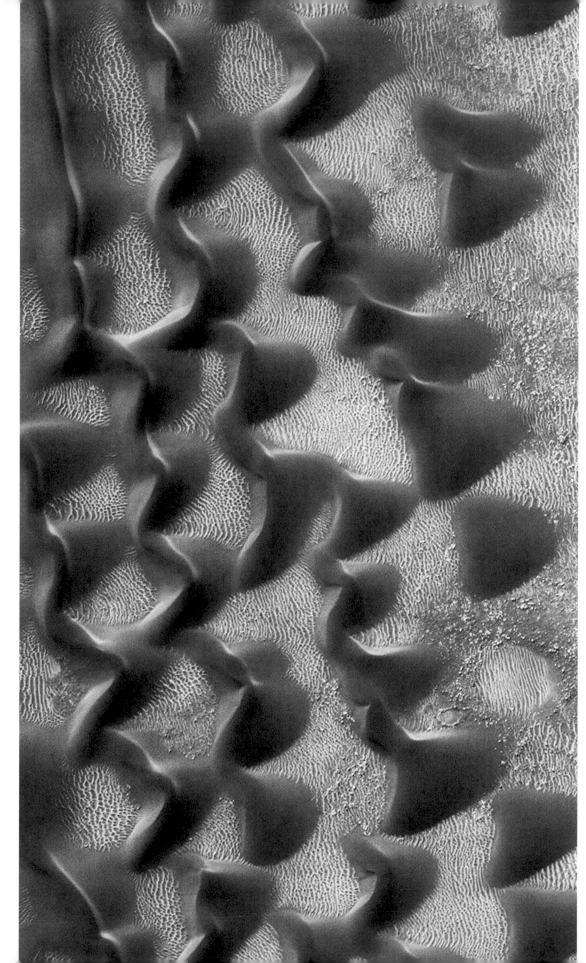

Michael Malin admits that many of his favorite images are featured on his website because they are 'just plain pretty'. Here is a field of dark sand dunes in the crater, Proctor, in the southern highlands. The dark material is probably fine-grained basaltic sand or dust; the lighter material between the dunes must have a different chemical composition.

But in the case of Mars we are working with one hand tied behind our back.

With no hammers, microscopes, or Geiger counters, geologists studying the stratigraphy of Mars have been forced to use a crude technique to divide the visible surface into different ages. As we have seen, the southern hemisphere appears to be the oldest. Not only does it contain the most craters, it also contains the largest. The planets are believed to have formed by the accretion of small bodies called planetesimals, and, once formed, to have continued to be bombarded by remaining bodies not yet swept up by the other newly formed planets. Very large impacts such as those that formed the Hellas and Argyre basins would have occurred early on in Martian history. As time progressed, the supply of large impactors would have started to dwindle. This is fortunate – if 100-km (60-mile) rocks were still routinely hurtling into Mars and the Earth, advanced life would never have stood a chance. Bacteria would undoubtedly survive the effects of a succession of impacts, but more complex forms of life would be repeatedly wiped out. The rate of cratering on Mars is not known, but we can make certain assumptions about the population of impactors in the inner solar system and by drawing analogies with the Moon, which inhabits roughly the same swath of orbital real estate.

The geological history of Mars has been divided by scientists into three broad time periods, or epochs. From oldest to youngest, these are Noachian, Hesperian, and Amazonian epochs, named after features on its surface. These epochs are defined purely by the number of meteorite impact craters visible on the surface, together with the presence of typical features such as volcanoes and lava plains. The Noachian extends from the birth of the planet around 4.5 billion years ago to between 3.5 and 3.8 billion years ago. Next is the Hesperian, which ended about 1.8 billion years ago, although uncertainty over cratering rates means that some scientists put the end of the Hesperian as early as 3.55 billion years ago. (Making definitive statements about *any* absolute ages of Martian surface features is extremely difficult.) Finally, the Amazonian extends to the present day. Amazonian features have few craters but show an astonishing variety. The Amazonian has seen the growth of Olympus Mons, extensive lava flows (some of which may be only a few hundred million years old), seas of sand dunes, and the frigid permafrost plains near the

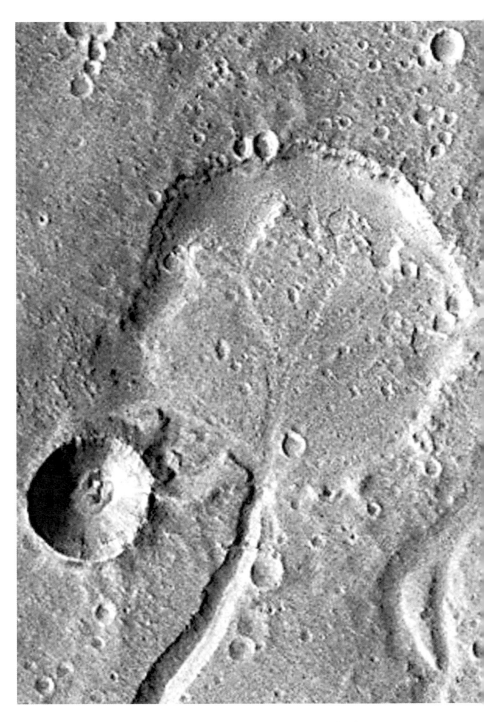

The existence of features like this delta-within-a-crater makes life difficult for those who argue that water has played no part in the formation of the Martian surface.

Martian poles. Note that most of the geological history of Mars – right up until the last third of the Amazonian, takes place during the terrestrial Precambrian. This alone shows just how different Mars and Earth are: the Martian surface has been preserved like Dorian Gray's picture in the attic whereas the Earth wears a thousand masks, each covering up the marks of the past.

Evidence for water flowing over the Martian surface throughout its long history is found almost everywhere, and throughout all three epochs. The presence of outflow channels – huge, scoured depressions tens of kilometers wide and hundreds of kilometers long – is a powerful argument, say most scientists, that a huge reservoir of water was present on Mars, perhaps beneath the surface, for much of its history.

Many planetary scientists interpret the available evidence to indicate that Mars has witnessed at least two periods during which liquid water existed on its surface for long periods of time. It is too cold, and the atmosphere too thin, for this to be possible today, but in the past it was different. The first of these wet periods was thought to be in the Noachian, around 4 billion years ago. Large valleys similar to those seen on Earth which can be seen winding their way across the surface of Mars date from this distant era. A second damp period, say supporters of the warm, wet hypothesis, happened more recently – between 1 and 3 billion years ago. For these features to have really been formed by water, Mars must have gained – and lost – an atmosphere thick enough and warm enough to allow liquid water to flow, on at least two occasions in its history.

Many of the Martian channels do not really look like rivers. What they resemble most is the aftermath of massive floods rather than the slow erosion of running water over thousands of years. These floods, if that is what they were, must have been enormous, in some cases far larger than the flow of Earth's largest rivers today. Other, smaller channels do resemble conventional river valleys, with branching networks of tributaries. Not all the 'water' features on Mars are channels. There are what look like lake deposits in many craters. And some have suggested that at one time much of the northern hemisphere of Mars was home to an immense sea, with the suggestion of a faint shoreline that can be seen to this day.

All this raises two main questions: if these channels were carved by water, where did this water come from? The sheer number of apparently fluvial features implies a reservoir of water of immense size throughout Martian history. And, just as importantly, where did this water go?

(*Opposite page*) Three gigantic Martian valleys: Dao Vallis runs diagonally from the upper left; Niger Vallis runs into it just above the center of the frame; Harmakhis Vallis is to their right. It is not known how old these valleys are, or how they were formed. Water is the prime candidate, but some geologists believe that other agents, such as carbon dioxide or even the wind, may have been responsible. The valleys are about 1 km (0.6 miles) deep and 15 km (10 miles) across.

(*Right*) All over Mars is evidence that tremendous forces once raged across its surface. These teardrop-shaped 'islands' in the Cerberus region look as though they were formed in a gigantic flood flowing from top right to bottom left.

Chapter 3
The Martian Eden

'Don't worry, Mars will not disappoint us'

Carl Sagan

When Mariner 4 sent its 22 grainy pictures back from its swoop past Mars in 1965, the world's planetary scientists breathed a collective sigh of despair. This was not Lowell's 'abode of life' (a phrase from the title of one of his three best-selling books), but a battered dust-world covered in craters like those that pepper the Moon. But the revelation by Mariner 9, which went into orbit around Mars in 1971, that Mars was covered in huge erosional channels forced a rethink. Some of Mars's surface was crater-free. Mars appears to have massive drainage basins which bearing at least a superficial relationship to those on Earth. Visual evidence of water is almost everywhere, from those enormous channels to the delicate gullies that some think are still active today. The Opportunity rover appears to have found evidence of an ancient shoreline at Sinus Meridiani, with ripples and sediments that can only have been deposited under water. And the chemical evidence indicates salty water. A shallow lake at least a few tens of centimeters deep must have existed here for several thousand years, even if it was covered by a layer of ice.

But today, Mars is extremely dry – at least on the surface. The only detectable water is the ice at the poles, plus a small quantity in the atmosphere. It has been calculated that if all this atmospheric water were to be precipitated, it would amount to a deluge of just a tenth of a millimeter (four-thousandths of an inch). An analysis of atmospheric gases carried out in the 1970s, when compared with the proportions of the gases in the Earth's atmosphere, suggested that Mars once had enough water to cover its surface to a depth of a few tens of meters. Losing this water – through the action of sunlight splitting the H_2O molecules and boiling the hydrogen off into space – would be easy. It could also have soaked into Mars's porous crust.

OCEANIS BOREALIS:
THE GREAT MARTIAN OCEAN

The channels tell a different story. Planetary geologist Michael Carr has calculated that to account for the amount of apparent fluvial erosion and deposition on Mars, the planet must once have contained enough water to flood the planet to nearly a kilometer deep. This would imply not only huge rivers bigger than the Nile, but lakes, seas, and even oceans. In the mid-1980s, Timothy Parker of JPL raised the possibility of a 'Boreal Ocean' covering the vast, low-lying, and antique northern plans, and claimed to have detected hints of an ancient shoreline in Viking imagery – in fact, several nested shorelines. This claim is controversial. Many scientists do not accept the existence of such a shoreline, but support has come from the Mars Orbital Laser Altimeter (MOLA) – an instrument aboard Mars Global Surveyor – which suggests that at least one of Parker's suspected shorelines lies at a constant elevation – compelling evidence for a northern ocean perhaps as big as the North Atlantic. In an article in *Sky & Telescope* in August

(*Opposite*) The north polar cap of Mars is a vast swirl of mostly water ice, surrounded by thousands of square kilometers of icy regolith. Although the polar caps were spotted as early as 1666 (by Giovanni Cassini), it was not until the 1780s that William Herschel argued that what we were seeing on Mars were fields of water ice – the same as at the poles on Earth. The composition of the polar caps has been a matter of debate ever since: were the caps made of water ice, frozen carbon dioxide, or both? It seems now that they are indeed mostly water ice, with a thin layer of CO_2 that 'burns off' in the summer. This MOLA image is color-coded to show altitude in relation to the Mars datum – blues and greens are low-altitude, progressing through browns, oranges and reds to white – the highest peaks.

2003, Stephen Clifford, a hydrologist from the Lunar and Planetary Institute in Houston, says he is convinced that not only did Mars once have an ocean, but this ocean covered a full one-third of the planet. He points out that young Mars would have had a much higher 'geothermal heat flow' than it does now (planets cool as they age as the radioactive elements in their interiors slowly decay).

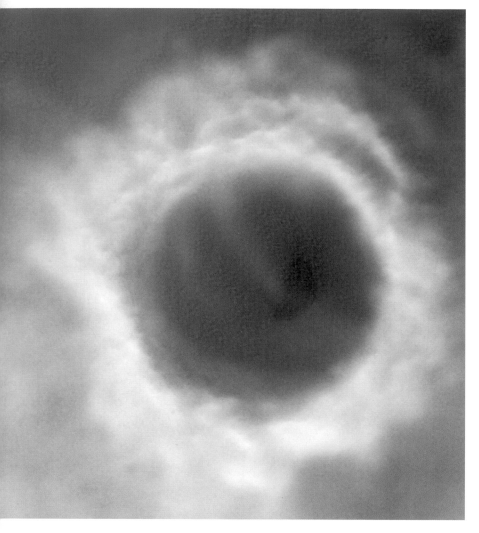

Like Earth, Mars is subject to four seasons thanks to its axial tilt of around 25°. This image, of the crater Lomonosov, was taken deep into the northern-hemisphere winter, on 20 April 2000, by the Mars Orbital Camera on board Mars Global Surveyor. Frost has highlighted the rim of the crater, and thin wispy clouds of ice crystals blur the view.

Clifford believes that the Martian ocean was a very early feature, which formed soon after the planet accreted and cooled some 4.5 billion years ago, long before the crust became cold enough to trap water underground in a gigantic permafrost layer. Later on, water stored deep underground in the southern highlands, which stand some 4 km (2.5 miles) above the northern plains, was released in episodically, causing the flood channels. By this time the ocean was either frozen over or had seeped into the crust, after which melting occurred only locally, when heat was released by a meteoritic impact or volcanic eruption. All the while, whatever water was in the upper atmosphere would be being dissociated by sunlight and leaking away.

In their 1991 paper in *Nature*, 'Ancient oceans, ice sheets and the hydrological cycle on Mars', Vic Baker and his colleagues went further, stating that there was good evidence that large parts of Mars had been flooded and desiccated on a regular basis, with an active hydrological cycle in place as recently as the Amazonian epoch:

> The channels nearly all debouched their floodwater to the northern plains of Mars, where ponded sediments, shoreline indicators, infilled craters and basins, spillover channels, pitted basin floors, whorled patterns of multiple ridges, and other evidence indicate the past evidence of extensive lakes or temporary regional flooding.

They go on to say that the Borealis ocean, which could cover an area of 40 million square kilometers (15 million square miles) with an average depth of 1,700 m (5,500 ft), repeatedly forms and dissipates as the climate swings between a temperate maritime state and the cold, dry climate with a rarified CO_2 atmosphere that we see today (and which, it would seem, is typical of conditions of Mars since its formation; the wet periods are atypical). Further support for the idea of an ancient ocean came in a paper presented at the 2004 Lunar

(*Opposite*) Ancient Mars? Trillions of liters of water lie beneath the Martian surface, locked away in a layer of permafrost that may be hundreds of meters thick. More water is frozen into the polar caps. If all this ice were to melt, it could cover much of the planet in a shallow ocean. This computer image shows what a truly warm, wet Mars may have looked like 4 billion years ago.

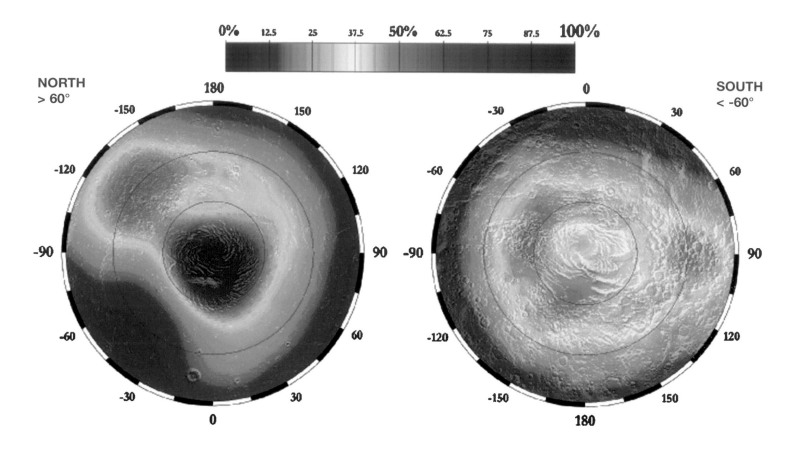

and Planetary Science Conference at Houston, Texas, in which S.C. Werner and Gerhard Neukum presented evidence that 'polygonal' terrains in the Utopia and Acidalia northern plains were formed on an ancient seafloor. These polygons consist of 200–800 m (650–2,500 ft) wide, steep-walled, and flat-floored troughs some tens of meters deep and 5–30 km (3–20 miles) in diameter. Crater-count analysis of these areas suggests 'extensive resurfacing effects' within an interval of approximately half a billion years – craters which would have formed between 3.8 and 3.4 billion years ago appear to have been somehow obliterated. This could have happened if these areas were under several hundred meters of water at the time.

Getting early Mars warm enough to be wet would have taken tremendous energy. Volcanoes are one solution; another is impacts from giant meteorites. Anthony Colaprete of NASA's Ames Research Center has suggested that impacts from space rocks at least 10 km (6 miles) across would vaporize all the ice buried in the target zone and throw a cloud of white-hot,

boiling rock into the atmosphere, heating the surface to hundreds of degrees, so melting and vaporizing even more ice. Eventually, things would cool down and all that vapor would turn to rain – and Mars would be subjected to cataclysmic flooding on a planetary scale. And then, after this brief and rather dramatic 'summer', Mars would once more revert to its cold and dry state.

If Mars really was once warm and wet, we need to know where all this water went. One possibility, of course, is the

(Above and opposite) One of the most important discoveries made about Mars in recent years has been that huge quantities of water ice are present in the Martian regolith (or soil), hundreds of kilometers from the poles. The neutron spectrometer aboard Mars Odyssey detected large quantities of hydrogen (shown here in orange and red) – and by inference, water – at low latitudes on Mars. This water would be a valuable resource for any future explorers.

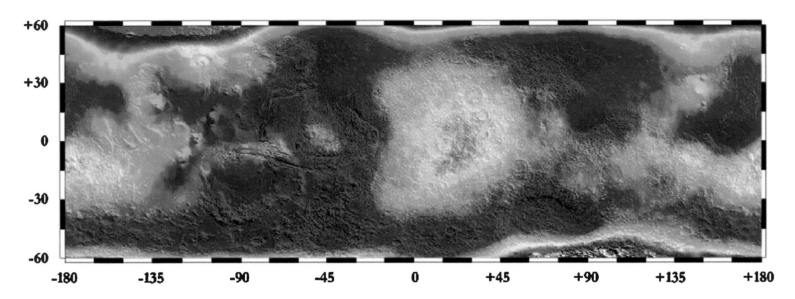

polar caps. After Cassini first saw these in 1666, it was later assumed (by analogy with the Earth) that they were made of water ice. Later, Ranyard and Stoney in Britain, eschewing the tempting terrestrial analogy, had argued in 1898 that the caps were a frost of frozen carbon dioxide. But then in 1949 Kuiper reported a spectroscopic study which left him certain that they were made of water ice. However, many spectroscopic observations from Earth suggested that dry ice rather than water made up the bulk of the caps' composition. The presence of large amounts of frozen CO_2 at the poles was confirmed in 1969, when Mariner 7 confirmed that the seasonal caps – the white areas that wax and wane in winter and in summer – are composed of dry ice, but the nature of the much smaller permanent caps was not confirmed. In 1976, the Viking orbiters measured the temperature of the caps accurately for the first time and found that the northern residual cap is far too warm to be composed of carbon dioxide; it must therefore be water. The southern residual cap was colder, but it is assumed that the dry ice covering is only superficial – as has now been confirmed by the Mars Express orbiter. Both caps together contain a substantial amount of ice, but not enough to account for all the drainage features seen on Mars, let

alone an ocean a kilometer deep or more in the northern hemisphere. In fact, if you melted all the ice at the Martian poles, you would get a planetary ocean only about 30 m (100 ft) deep.

There are many theories about what would have happened to the water in the northern oceans (which, it is predicted, had a volume of about 23 million cubic kilometers, or 5.5. million cubic miles). Michael Carr and Jim Head have suggested that some of it, about 30 percent, could have been lost to space by sublimation (vaporizing directly from the solid) followed by dissociation (the breakup of the water molecule into free hydrogen and oxygen by the action of solar radiation), almost 20 percent could be in the present polar caps, and the rest could be trapped in other volatile-rich surface deposits or redistributed in a subsurface groundwater system.

In 2002, there was much excitement when the Mars Odyssey spacecraft's neutron spectrometer found the distinctive signature of hydrogen from huge areas near the Martian equator. As far as the neutron spectrometrists are concerned, hydrogen equals water; it cannot exist as a free gas, as it would quickly float off into space. Explaining the presence of huge quantities of hydrogen compounds other than water is difficult. The spectrometer *must* therefore have found huge quantities of water ice in the top few meters of Mars's soil. It is not clear just how deep this water extends, but the assumption is that at least the top 100 m (300 ft) or so is effectively permafrost. This is a massive amount of water.

The first reports of the new water finding came in March 2002, although more publicity was attracted with the publication of a paper in *Science* in late May of that year. The lead scientist behind the finding was William Boynton, of the University of Arizona. At the time, he said that, 'If this is confirmed, it is fantastic. There is the equivalent of at least several percent water south of 60 degrees latitude.' In the *Science* paper, the authors noted that:

> It is important to note that we find a strong subsurface hydrogen signal on Mars essentially everywhere that ice is expected to be stable and where our signal is not obscured by CO_2. The total pore space in the regolith has been estimated to be sufficient to contain ice equivalent to a global water layer 0.5 to 1.5 km [0.3–1.0 mile] deep. Our results, of course, do not reveal anything about whether or not ice is present in the enormous volume of regolith that lies below the roughly 1-meter [3 ft] depth to which the gamma and neutron techniques can sense. However, they certainly are consistent with the view that the subsurface regolith may be a substantial reservoir for Martian water.

The news of the Martian ice made many newspapers and the possibility of life in the distant past was raised, and many pundits pointed out that a large reservoir of near-surface water (it was calculated that a bucket of Martian soil, warmed up, would contain at least a liter of the stuff) meant that Mars was eminently habitable. Not only would there be plenty of water for astronauts to drink, there would also be a ready supply of rocket fuel on the surface. Water is composed of hydrogen and oxygen. It is (fairly) straightforward to split these two chemicals apart, using solar-powered electrolysis, creating a supply of the two gases which can be recombined in a rocket's combustion chamber. A crewed expedition to Mars would not only be able to slake its thirst, but would also be relieved of the burden of carrying all the fuel necessary for the trip home all the way to the Red Planet. With several quadrillion tonnes of ice buried beneath the surface,

Mars changed once again. Not so harsh, not so hostile. A Mars covered in ice becomes a softer, damper world. Not only would human visitation be possible, but human *habitation* – perhaps permanently. With all that water you could till the soil (though you would need to find some way of protecting the plants from ultraviolet radiation). If only there were some way of making Mars warm again, it could be made wetter as well. Perhaps, in centuries hence, the great craters and basins will once again start to fill as water rains down from an orange sky. Argyre and Hellas will become small seas, the Valles Marineris will fill, and water will cascade down from the southern highlands to fill up slowly a dead ocean basin, whose coasts once lapped for a near eternity with the gentle wash of tideless Martian waves. Lowell's canal-builders may be dead, but they would be resurrected, this time in human form. The subsurface ice is bringing Mars back to life.

GAIA ON MARS

One of the most enthusiastic proponents of this viewpoint is John Brandenburg of the Florida Space Institute. Brandenburg is not a member of the Mars mainstream. *Dead Mars, Dying Earth*, a book he wrote with Monica Rix Paxson, claims that there is good evidence of archaeological remains on Mars, particularly in the Cydonia region, home to the infamous 'face'. Nevertheless, Brandenburg's ideas about Martian climate change can be taken seriously as they represent merely one end of the spectrum that has the warm, wet Mars at one end and the cold, dry White Mars at the other.

Brandenburg believes that a large ocean existed until 'probably the early Amazonian'. He believes, furthermore, that the relative scarcity of craters on the northern plains points to a relatively young age for this ocean – far younger

(*Opposite*) Kees Veenenbos is a Netherlands-based computer artist and Mars enthusiast. Unlike many renderings of 'alternative' Martian scenes, his images are not fantasies – they are based on hard data (from the Mars Orbital Laser Altimeter aboard the Mars Global Surveyor spacecraft), combined with computer-imaging software that is able to show the planet as it might have been under different climatic conditions. Here, the crater Gusev (the landing site for the Spirit rover in January 2004) is shown partly filled with water – which it may have been more than 3.5 billion years ago.

than supposed by Stephen Clifford. He even gives this ocean a name – the 'Malacandrian Ocean', which

> would have had a depth of several kilometers, covered one-quarter of the planet, and contained an approximately 400-metre-deep [1300 ft] planetary-wide layer of water. It would have played an important role as a thermal and atmospheric buffer to support and stabilize the Martian climate system.

I ran into Brandenburg at the 2003 Mars meeting in Pasadena. He thinks that not only do we need to reassess the whole concept of Mars as an essentially cold, lifeless place, but we need to accept that it was once almost certainly very much alive. 'Basically, Mars is a very frustrating puzzle for us now,' he says

> Mars is red because it is highly oxidized. It is funny, but scientists just don't seem to want to talk about the fact that Mars is oxidized. This means that at one time in the past Mars must have had a lot of oxygen around.

The source of this oxygen is indeed poorly understood. The fact that Mars is 'rusty' is seen by people as proof that Mars was once a very Earthlike place indeed. To Brandenburg, the answer to why Mars is red is obvious – all the oxygen in the atmosphere somehow got incorporated into the rocks. But the problem for the supporters of the warm wet hypothesis is explaining how and why Mars has changed so dramatically. The problem is simple to state, hard to solve. It goes like this: Mars is a small planet, less than one-sixth of the volume of Earth. It has low surface gravity, about a third that of our planet, which means that atmospheres slip away. On Earth, light gases such as hydrogen quickly float off into space. Once we had plenty of hydrogen in our atmosphere, but now it has almost all gone. On Mars, even heavier gases such as nitrogen find it easy to make their way into the void.

Mars's smallness affects its climate in other ways too. Little objects have a greater ratio of surface area to volume, so they cool off more rapidly than big ones. Not only would Mars have cooled far more rapidly than Earth and Venus (which are comparable in size) from its fiery birth, but it has also been leaking heat away from its interior at a much greater rate ever since. The insides of rocky planets are hot because of the heat generated by the decay of radioactive isotopes in the rocks, gravitational compression, and the energy left over from their formation. It is estimated that the amount of heat escaping from the interior is no more than 40 percent of the heat lost similarly from Earth. Because Mars is small, its crust solidified quickly, and solidified to a great depth. Although Mars has some impressive volcanoes it totally lacks the great chains of fire that mark the boundaries of Earth's global system of plate tectonics. Mars's crust solidified too quickly and too thoroughly for convection currents in the planet's hot mantle to break it up into plates as happened on Earth. On our planet, plate tectonics is an efficient way for the planet to shed heat.

Second, and following from this, Mars is very, very cold. It is farther than Earth is from the Sun and receives only 43 percent of the solar radiation that we do. Mean average surface temperatures today range from –55°C (–67°F) at the equator to about –119°C (–182°F) at the poles. The coldest place on Earth – the high Antarctic plateau – is far warmer than the warmest place on Mars. While it is true that temperatures in the bright sunshine of a Martian equatorial summer's day do occasionally creep near to, and even above, the freezing point of water, for most of the time most of Mars is cryogenically cold. Temperatures on its surface are more like those on the surface of the moons of Jupiter or on Saturn's Titan than those on Earth.

Now, if Mars was once warm and wet, some of these conditions must have once been different. Dramatically so. But which ones? Mars has not shrunk since it was formed. Mars would have had no more gravity and have been no more capable of hanging onto an atmosphere 3.5 billion years ago than today. Maybe the Sun was warmer back then? Unfortunately, the opposite is true. During the early Noachian epoch, the solar output was probably no more than three-quarters of its output today. Mars would have been receiving even less heat from the Sun than it does now. If the Mars of today was transported back in time 4 billion years then, all else being equal, its surface would be even more frigid than it is today.

So all things cannot have been equal, if Mars was once the watery paradise that Brandenburg envisages. His answer is simple: Mars must have once had a thick atmosphere of carbon dioxide – quite enormously thick, maybe giving it a

Viking images of Hecates Tholus, a 5,300-m (17,400-ft) high volcano, show numerous craters which may be volcanic in origin rather than caused by impacts. Until quite recently some astronomers believed that most if not all craters on the Moon and Mars were volcanic; the fact that they were caused by impacts has only really been accepted since the days of Apollo. Also in this picture can be seen lines radiating outwards (and downhill) from the central caldera. These may be channels, carved by water.

mean surface pressure of several bars. Carbon dioxide is a heavy gas, heavy enough not to escape even from Mars's feeble gravity. In addition, large amounts of geothermal heat plus the greenhouse gases, water vapor, methane, and sulfur dioxide would have turned old Mars into a pressure cooker.

Brandenburg insists that if you give Mars a thick enough atmosphere, you can just about get temperatures high enough for liquid water to be stable. There is another problem, however: a thick CO_2 atmosphere combined with an ocean

This rendering by Kees Veenenbos shows the Shalbatana Vallis in ancient times, being carved by the mighty torrential floods coming from the Valles Marineris.

would be chemically unstable. The carbon dioxide would simply dissolve in the water, forming seltzer, or carbonic acid, which would then combine with silicate rocks to form carbonates – limestone. In this way, any thick CO_2 blanket would effectively be turned into rock, and pressures (and temperatures) would rapidly fall. The long-lived ocean Brandenburg proposes would be impossible. (Venus avoids turning its CO_2 into limestone, as happens on Earth, because there isn't any surface water, although this may not have been the case in the distant past.) So there must be some sort of carbon recycling process going on. One possibility, Brandenburg notes, is that Mars may once have had a very limited plate tectonics regime and much active volcanism. Periodic volcanic flooding of large areas of Mars, combined with the subduction of crustal plates into the mantle, would have allowed the planet's internal heat to 'cook' the limestones and release CO_2 back into the atmosphere. Another possibility was raised by Jeffrey Moore, of NASA Ames, in a paper published in *Nature* in April 2004. Recent findings from NASA's latest Mars landers suggest that sulfate minerals rather than carbonates may have been the 'dominant aqueous minerals on Mars'. This suggests that Mars's early 'warm, wet' climate may have been 'propped up' by sulfur dioxide. The Martian waters would have been dilute sulfuric acid, and this would chemically prevent the formation of carbonates. Only once enough sulfur dioxide had combined with surface minerals to form sulfates could the predominantly CO_2 atmosphere 'collapse'.

But, according to Brandenburg, there is a more powerful geochemical engine for keeping the atmosphere stable, he maintains: biology. In what he calls his New Mars Synthesis, he paints a novel picture of the Red Planet. Instead of a deeply alien world, which in the most fundamental ways has differed from Earth since the two worlds were born, he thinks that Mars and Earth were similar for maybe as much as 4 billion years. If true, then Mars has been a cold, dead desert for less than half a billion years or so, if not quite the blink of an eye then certainly a mere forty winks in geological time. And he maintains that there is no reason why Martian and terrestrial evolution were not proceeding in similar – or at least analogous – ways, all through this vast epoch. That means that life did not just evolve on Mars, it thrived. Forget microbes, cyanobacteria and primitive algae. Brandenburg thinks that Mars may have been home to animals, possibly even intel-

Olympus Mons – the Martian Mount Olympus – is the Solar System's most impressive mountain, towering 21,000 m (69,000 ft) high (two and a half times the height of Everest). This rendering depicts the area as it might have been three billion years ago, with the great volcanoes covered in ice and snow, towering above the watery plains.

ligent life. Five hundred million years ago, life on Earth was just starting to crawl out of the ocean. A hundred million years before, an explosion of evolutionary ingenuity had created the basic body plans for the flora and fauna that live here to this day. The seas of Cambrian and Silurian Earth teemed with exotic creatures – trilobites and armored fish, giant underwater arthropods, and huge worms. Might Mars have been the same? If so, then when astronaut-geologists get down there with their hammers, they may find some rather more interesting fossils than a few stromatolites and bands of algae.

This life is the key to Brandenburg's vision. Life would have maintained the equilibrium of the atmosphere by photosynthesizing CO_2, and also would have formed a protective varnish on Martian rocks to prevent excessive carbonate formation. Life is also good at making methane, one of the most powerful greenhouse gases. Life is the main engine behind a life-friendly Mars, hence the 'Martian Gaia' title to his hypothesis.

So, could he be right? And what is his evidence that life did evolve in this way? Well, he maintains that the highly oxidized surface of Mars is evidence in itself. Water in Martian meteorites, he maintains, suggest that Mars's surface was once wet. But the most compelling evidence for ancient Martian life is simply the presence of an ancient Martian ocean: 'For this ocean to be stable, you needed a biosphere. A massive biosphere can stabilize the oxygen on the surface, just as it does on Earth.' Brandenburg also points to the latest evidence from the Mars rovers. The sulfates they have discovered, he says, could have been made by living organisms.

But Mars is clearly not like this now; something happened to bring this Eden-like state to a rapid end. In Brandenburg's view, something happened to Mars half a billion years ago to turn a warm, soggy and fecund planet into the arid hellhole we see today. Brandenburg points to a possible smoking gun for what killed Mars: 'The Martian dependence on a heavy CO_2 greenhouse also allows the possibility for catastrophic climate change because of the bistable nature of such a greenhouse on Mars.' In other words, if something dramatic happens, then the warm, wet Martian climate can suddenly flip. But what could have caused this catastrophe? 'Life created an environment where it could sustain itself', he says. 'You have a Mars that was managing to preserve its environment. Then, to get the Mars we see today, you had to end it. How did that happen?'

THE FALL OF MARS

There are several candidates for what Brandenburg has christened the 'fall of Mars'. One is a truly cataclysmic volcanic eruption, or series of eruptions. A huge explosion in the Tharsis chain could have thrown trillions of tonnes of dust into the atmosphere, blocking off sunlight and initiating global cooling. It is known that major volcanic events on Earth, such as the huge outpourings of basalt in Siberia during the Permian, can disrupt the whole planet's climate enough to cause at least a minor extinction event. Even modern isolated volcanoes like Mt Pinatubo in the Philippines, which erupted dramatically in

'Evidence for a massive photosynthetic biosphere will be found on Mars, such as coal and petroleum deposits'

John Brandenburg

1991, have a measurable (albeit temporary) cooling effect. The other possibility, according to Brandenburg, is a particularly nasty asteroid strike. 'This would lead', he says, 'to the Martian Chicxulub', alluding to the catastrophe here on Earth that is thought to have seen off the dinosaurs and many other species 65 million years ago. He even thinks he knows where and when this impact occurred. 'There is evidence that the formation of the Lyot impact basin in the Early Amazonian, a 200-km [125-mile] double-ring crater, may have triggered the collapse of the Martian climate. Mars fell and could not get back up again.'

So how did a meteorite strike plunge Mars into deadness? Why was Mars so vulnerable? After all, we know that the Earth has been hit many times (presumably more often than Mars, given that it is a much bigger target), and life has continued nonetheless. Brandenburg explains:

As evidenced by the longevity of the paleo-ocean, Mars's climate system ran smoothly for a long geological period, apparently until the early Amazonian, then, this catastrophe occurred. A heavy CO_2 greenhouse on Mars is unstable due to the fact that temperatures on Mars can easily dip so low that carbon dioxide can condense on the planet's surface. This means ancient Mars had two stable atmospheric states, state one with a warm, dense atmosphere trapping lots of heat, and state two with a thin, cold atmosphere with most of the gas frozen onto the surface. The transition between the two states could require merely a large temporary thermal excursion. For a Mars with a heavy greenhouse, even with a large ocean to act as a buffer, catastrophe was just one large chilling event away. Such a catastrophe would push the temperature at the poles below the point where dry ice would form, leading to collapse of the atmosphere onto the poles, loss of the greenhouse effect, thus leading to more cooling. The ocean surface would freeze, decoupling it from the atmosphere and leading to an even more rapid decline in temperature and pressure.

According to controversialist John Brandenburg, warm and wet conditions persisted on Mars until quite recently – perhaps as recently as half a billion years ago. This would, he argues, have given Mars enough time to evolve a biosphere on a par with the Earth's. Bringing this 'Martian Eden' to an end could have been a particularly catastrophic impact, and he has identified the 132-km (82-mile) crater Lyot as a possible 'smoking gun'. Brandenburg is well out of the Mars mainstream – but in 1966, say, anyone suggesting that Mars may have once had flowing water would have been dismissed as a fantasist.

years ago) something triggered a global ice age, causing the oceans to freeze over all the way to the equator. The snowball Earth was brilliant white, and, in a catastrophic runaway feedback mechanism, temperatures plummeted even lower, threatening to freeze the oceans to their beds. Fortunately, carbon dioxide saved the day. The planet may have been covered in ice, but that didn't stop Earth's muscular volcanism, which continued to pour CO_2 into the atmosphere. With no liquid water in which to dissolve, CO_2 concentrations in the atmosphere rose dramatically, eventually triggering a massive greenhouse effect that not only melted the ice in short order but probably threatened not to stop there and go on to make our world another Venus. Again, life, which had been hanging on in isolated geothermal pools, on the seafloors and around hot springs, saved the planet for posterity by modulating the CO_2 cycle.

There is good evidence for the snowball Earth – namely the presence of glacial deposits in rocks that were known to have existed in the tropics, and of a distinct iron-rich layer that could have been formed at the time of the spectacular remelt, when the carbonic oceans dissolved rock by the cubic kilometre. But what evidence do we have for such a catastrophe on Mars? According to Brandenburg, one of the most compelling arguments is the simplest and most obvious of all – Mars's very redness. But he maintains that when humans (or at least very advanced robots) go to Mars, his theory will be vindicated.

Brandenburg's hypothesis is testable, he claims. He predicts that if it is correct there will be large amounts of water ice around the northern pole. This ice will melt at its base, thanks to geothermal heat, forcing carbonated water into the Martian soil. He predicts that large carbonate beds should thus be found at the poles, as well as large deposits of silica. Secondly, the redness of Mars will be found to be due to Earthlike weathering of lavas in an oxygen- and CO_2-rich atmosphere. 'Evidence for a massive photosynthetic biosphere will be found on Mars, such as coal and petroleum deposits.' And third, Brandenburg is sure that

So the asteroid hit, dust was thrown into the atmosphere, blocking the sunlight, and Mars plunged into a winter from which it has not recovered. Some scientists believe that something similar nearly happened on Earth, coincidentally at about the same time that Mars was dying. The 'snowball Earth' hypothesis, which started off as rather left-field but which has gained a surprising amount of credibility among geologists in recent years, states that around the beginning of the Cambrian (some 500 million

evidence for evolved biology on Mars will be found, given the longevity of the biosphere. These would include, by Earth analogy, perhaps considerable advances over the primitive microbes associated with early planetary environments. There may be surviving pockets of life on Mars, bacteria, who knows, maybe even blind salamanders.

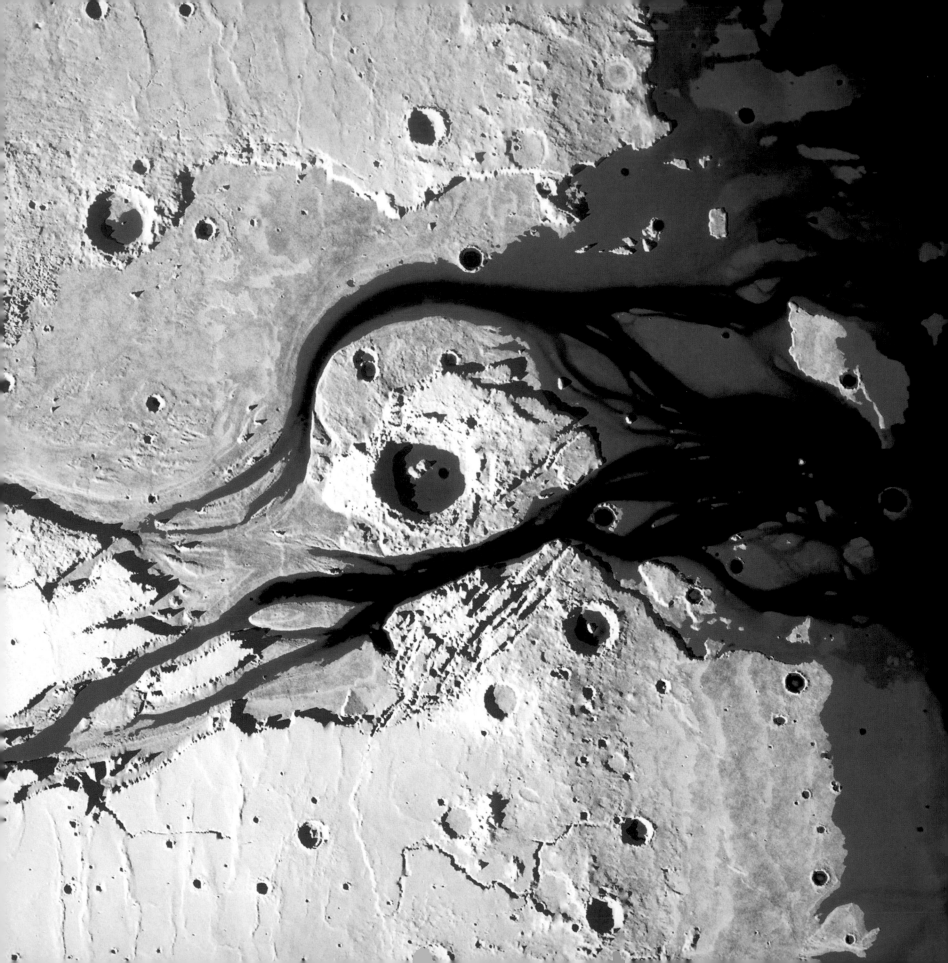

'Carl Sagan, in response to concerns over whether Mars was going to be boring, once said, "Don't worry, Mars will not disappoint us"', Brandenburg comments. 'This is certainly true, for after over a thousand years of human longing for more information about Mars, and almost half a century of space probe visits, the tale of Mars is not near being told.' Brandenburg's New Mars Synthesis represents one end of the spectrum that is today's thinking on Mars. His views are extreme, and are certainly not shared by most scientists. Few people you speak to will talk about salamanders and reptiles living on Mars half a billion years ago, and almost all of Mars scientists shudder when they hear talk of 'the Face'. Yet, in its essence, his ideas are not that far off the mainstream. Most researchers believe that Mars was once more Earthlike than it is today. There is certainly evidence – the water channels, the tantalizing signs of an ocean, and those Meridiani sediments. People clearly want to believe that Mars was once warm and wet; that it had life, great blue rivers, lashing rainstorms and glittering oceans; microbes and weeds, possibly fish – who knows? Maybe Mars was like this once. But it is not the only possibility.

(*Opposite*) This Kees Veenenbos rendering shows a perpendicular view of part of the crater Sharonov, which lies towards the southern outlet of the Kasei Valles channel. Some geologists argue that through this channel may once have gushed more water than flows through the Earth's largest rivers.

Chapter 4
A Cold, Dry Place

'There is a lot of funding in wet, blue Mars. It is part of the human psyche, isn't it, to want to not be alone?'

Nick Hoffman (2003)

Not everyone is convinced that Mars is – or even ever was – quite as wet as most now think. Although with his views he is very much out on a limb, Nick Hoffman, perhaps the leading skeptic of the warm, wet hypothesis, is a respected dissident in the Mars community. He is a geologist at the University of Melbourne and, like Michael Carr (perhaps the leading authority on Martian water), a Briton by birth. Hoffman believes that much of the modern thinking about Mars is wrong. He believes that not only does Mars have very little liquid water on its surface today – if it has any at all – but that liquid water has played a marginal role in the evolution of the Red Planet. He believes that there never was a Martian ocean, and that the great craters and basins such as Hellas and Argyre were never glittering blue lakes. Hoffman also thinks that many of the huge drainage channels were possibly carved not by water but by another substance altogether – carbon dioxide gas. Furthermore, he sees the gullies as evidence that pressurized gaseous carbon dioxide, not water, is the dominant erosional agent (together with wind) operating on Mars today.

If Hoffman is right, then Mars is far from the fallen Eden that some scientists believe it to be. For a start, the chances of life without long-standing surface water would be very much diminished. Hoffman's Mars is probably sterile, at least on the surface. Hoffman's Mars – White Mars, he calls it – is anathema to people who, like John Brandenburg, believe in a wet Mars. Hoffman's Mars, cold, lifeless, frigid, and with an alien gas as its dominant agent of erosion, is much more unearthly than current thinking suggests. But, he says, this is precisely the point:

Scott wrote of the Antarctic, 'Dear God, this is an awful place.' And Mars is the proverbial awful place. We want to think of Mars and being quite like the Earth, a bit colder and with less air, but it is not. It is truly inhospitable. An awful place indeed. I think that instead of thinking of it as a cold version of the Earth, we should instead think of it as a warm version of Pluto, or Europa, like an icy moon from the outer solar system.

Hoffman believes that Mars differs in fundamental ways from the 'true' terrestrial planets of the inner solar system, Earth, Venus and Mercury. Possessing a thin atmosphere of carbon dioxide, and significant quantities of the gas, either in solid or liquid form underground, Mars resembles in several key aspects the large frozen satellites of Jupiter and Saturn. Liquid water simply has no role to play. It is too cold, and always has been.

Perhaps Nick Hoffman's fiercest critic is Vic Baker. 'Perhaps he is right,' he says,

but if so then Mars is Planet Bizzaro. It looks like it had water, it has features that resemble features we know on Earth were made by water, so why can't they have been made by water on Mars? Why invoke some weird mechanism about which we understand so little?

(*Opposite*) Sunrise over Olympus Mons. This Veenenbos rendering shows Mars in its current state – no flowing water, no creeping glaciers.

Hoffman's reply is that in planetary science it is unwise to make assumptions. In particular, we should not assume that because something on Mars looks like something on Earth, then they must have been made in the same way – a lesson that the photogeologists learned about the Moon during Apollo:

> People say that because there are features that look like they were made by water, at first glance, then why not assume they were made by water? After all, if it has a beak like a duck, walks like a duck, and looks like a duck, then why not assume that it is, most probably, a duck? But I would counter this with the duck-billed platypus. It has a beak like a duck, swims, and lays eggs. Yet it is most definitely not a duck.

'if you are going to spend $30 billion designing and flying spacecraft to Mars, it might be worth doing a bit of preliminary work first.'

Nick Hoffman

The current Mars exploration program is predicated not only on the search for water, but also on the search for life – either ancient or extant. The latest NASA Mars probes, the Mars Exploration Rovers, arrived in January 2004. NASA's follow-the-water strategy is clearly setting the course for its Martian exploration program, and this, says Hoffman, may have been an expensive mistake. He believes it may be better to try to do some simple (and relatively cheap) work in the lab first before rushing off to look for Martian waterfalls:

> If water is responsible for the floods on Mars, then this fact is fundamentally important to the question of whether there was once life on Mars. If water is found, today, in a warm layer, then there is every possibility that life would have evolved and maybe still exists. Water on Mars is a major step towards assuming life. This has profound scientific and philosophical consequences. If there is an alternative explanation that does not involve water, this would not have these profound implications. So we have to do work on our models in the lab and prove it one way or another before

we send all these hideously expensive experiments to the place to look for life. In other words, if you are going to spend $30 billion designing and flying spacecraft to Mars, it might be worth doing a bit of preliminary work first.

NICK HOFFMAN'S MARS

Carbon dioxide, not water, shapes the surface of Hoffman's Mars:

> My White Mars model explains all the existing problems about Mars's surface, atmosphere and evolution. It involves only one unusual aspect – that the active fluid agent for those outburst 'floods was the *other* volatile – CO_2. The story of CO_2 on Mars is as rich and complex as the story of water on Earth.

Hoffman's theme is that Mars is not boring because its surface has not been sculpted by water; instead, Mars is interesting precisely because its surface was sculpted largely by something that is *not* water. We understand much about how water shapes the land, how it etches into hills and washes sediments downstream. On Earth, water is all-pervasive, the key to deciphering just about every landscape you can imagine. Even our driest deserts are sculpted by water. But on Mars, although we see plenty of evidence for active erosion and deposition, Hoffman believes that the processes operating are fundamentally alien.

His starting points are a number of what he sees as paradoxes that simply cannot be explained by the warm, wet model. The most serious problem is the 'young, cool Sun' paradox. When it was realized that our star's energy output 3 billion years ago was a mere 70 percent or so of what it is now, calculations suggested that average surface temperatures on Earth should have hovered around the –15°C (5°F) mark. Yet there is no geological evidence that the very early Earth was ever that cold (even during the putative snowball Earth era, billions of years

(*Opposite*) The dry-ice vistas of the Martian south polar cap are probably the most alien on the planet. Here the dominant process shaping the landscape is the sublimation and condensation of hundreds of cubic kilometers of frozen carbon dioxide, producing a surreal pattern of swirling pits, cliffs, dry-ice mesas and buttes.

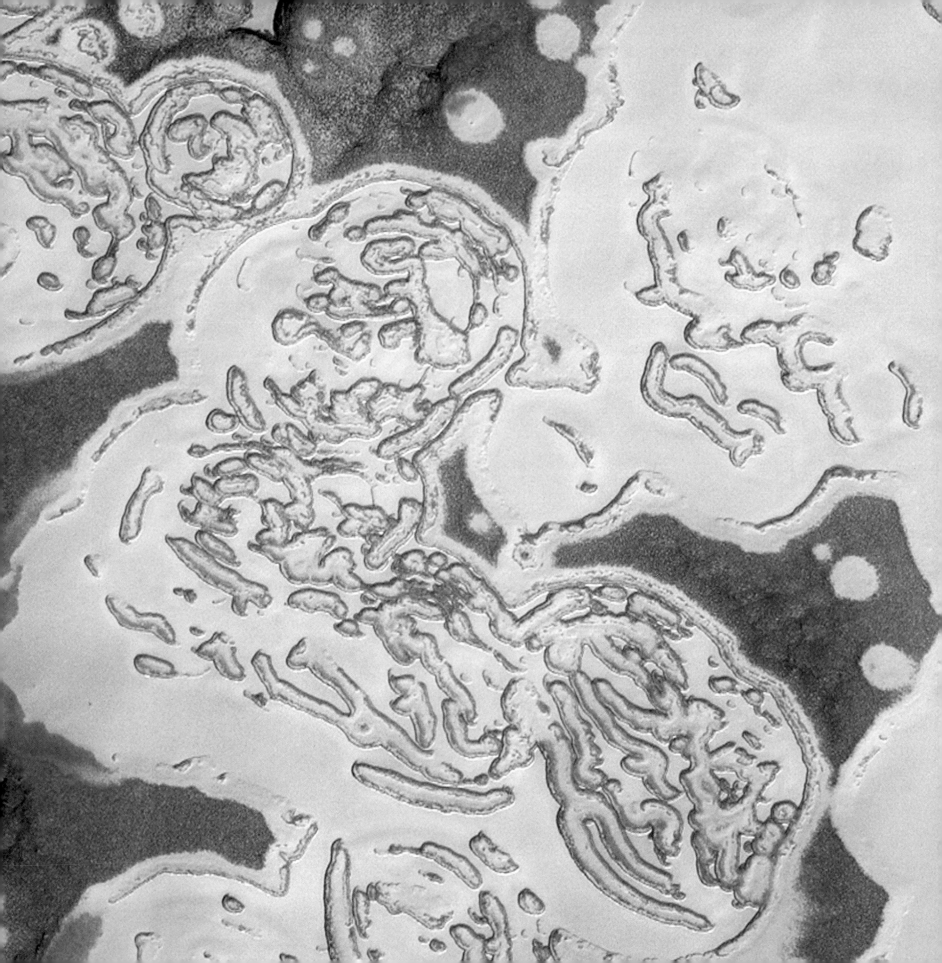

later, temperatures never dipped that low, and there is good evidence that there was liquid water on its surface within 100 million years of our planet's formation). The resolution of the paradox on Earth was the greenhouse effect. This means that not only was the young Earth warm enough to keep water liquid and get life started, it is also warm enough for us to live on it today. Mars gets only 43 percent of the solar energy input received by Earth. Venus, on the other hand, gets 193 percent of what we do (the energy received by a planet orbiting a star falls off with the square of the distance from the Sun). This means that Venus is more like Earth than Earth is like Mars when it comes to solar heating.

The problem, then, is to make an early Mars, which would have received even less heat than it does today, warm enough for water to flow. And you also need to invoke a mechanism to stop Earth from overheating and turning into Venus. According to Hoffman, this has led to a certain amount of scientific gymnastics, as atmospheres are tweaked and twiddled in order to get the desired effect:

As far as we can tell, all the terrestrial planets received broadly similar volatile inventories and accreted in similar fashions (although the late-stage impact that formed the Moon has made Earth a very special planet in several ways). The question of the evolution of the atmospheres and surfaces of the terrestrial planets needs to be treated as a single question, not as three separate ones.

Over the last couple of decades, two threads have permeated atmospheric studies of the terrestrial planets. One is the desire to make early Mars warm and wet, by using special greenhouse cocktails and modeling clouds and atmospheric properties to keep it warm. The other is the desire to avoid a runaway greenhouse on the early Earth, by using special greenhouse cocktails and modeling clouds and atmospheric properties to keep it cool. Obviously, these are completely contradictory goals.

According to the warm, wet Mars people, Mars had simply a thick, soupy CO_2 atmosphere to keep it hot. The trouble is, says Hoffman, this simply doesn't work. If Mars once had a thick atmosphere, why does it not have one now? The traditional explanation is that the atmosphere was blasted off by meteorite impacts or by the Solar wind, but if that

happened, then why didn't Mars also lose, say, its water (which in a Greenhouse Gaia would have been liquid and vapor)? Mars clearly did not lose all of its water, as trillions of tonnes of ice have recently been discovered permeating the topsoil. Hoffman believes that Mars *never* had a thick atmosphere, except for short bursts of time after a large impact generated enough heat to melt/vaporize a lot of water and carbon dioxide. Such an atmosphere would have been stable only 'on a timescale of hours or days, months possibly', he maintains. The early, Noachian Mars would have had surface temperatures even lower than they are today, and pressures would have been lower as well. Instead of being warm and wet, the young Mars would have been 20 degrees colder and had an atmosphere just one-seventh the pressure of what we see now. Not warm and wet, but very, very cold and extremely dry.

Also, and this is perhaps the most serious objection to the warm-and-wet hypothesis, there is the almost complete absence of limestone on the Martian surface. If Mars was once covered by substantial amounts of liquid water, for substantial amounts of time, and had a thick CO_2 atmosphere, there *should* be huge deposits of carbonate minerals on its surface. This is because of these minerals' unique relationship with water. Carbonates, such as calcite, magnesite, and dolomite, form when carbon dioxide gas from the atmosphere dissolves in water. If Mars was once warm (because it had a thick carbon dioxide atmosphere) and covered with water – including a large northern ocean – then its surface should be covered with limestone.

As it turns out, there *are* some carbonates on Mars. In August 2003, Arizona State University's Joshua Bandfield, Phil Christensen, and Timothy Glotch published a paper in *Science* detailing how small amounts of carbonate minerals – around 2–3 percent – had been found in several locations across the Martian surface by the Thermal Emissions Spectrometer (TES) aboard the orbiting Mars Global Surveyor. But the quantities of carbonate material were tiny – and, most importantly, they were not in the form of large rock formations. In fact, the TES detected the mineral's presence in Martian dust in quantities of between 2 and 5 percent. 'We have finally found carbonate, but we've only found trace amounts in dust, not in the form of outcroppings as originally suspected', wrote Christensen.

Tiny amounts of water in Mars's atmosphere can interact with the ubiquitous dust to form the small amounts of carbonate that we see. This seems to be the result of a thin atmosphere interacting with dust, not oceans interacting with the big, thick atmosphere that many people have thought once existed there.

The finding is ammunition for those who suspect that Mars never has been really warm – or wet. 'This really points to a cold, frozen, icy Mars that has always been that way, as opposed to a warm, humid, ocean-bearing Mars sometime in the past', says Christensen. 'People have argued that early in Mars history, maybe the climate was warmer and oceans may have formed and produced extensive carbonate rock layers. If that was the case, the rocks formed in those purported oceans should be somewhere.' Although ancient carbonate rock deposits might have been buried by later layers of dust, Christensen points out that the global survey found no strong carbonate signatures anywhere on the planet, despite clear evidence of geological processes that have exposed ancient rocks.

In a news article accompanying the paper, *Science* staffer Richard Kerr wrote:

On Earth, water and atmospheric carbon dioxide combine to form carbonic acid that eats away at rocks and flushes their remains into lakes and oceans. There, dissolved carbonates will eventually precipitate to form solid carbonate deposits like the White Cliffs of Dover, with or without the help of animals that form carbonate skeletons. But geochemists imagined that wet weathering would produce something like 20% carbonate in Martian dust, not 2%.

The spectral data pouring back the last few years from instruments orbiting Mars are starting to build a case for a volcanic planet only gently touched by the chemical alteration that water has wreaked on Earth for billions of years. In the emerging view, a perpetually frigid climate has kept Mars's modest store of water locked up as ice most of the time. The surface may never have been particularly hospitable to life, at least not for long. If so, the search for signs of life will have to focus on the Martian interior and its rare surface exposures.

There are no White Cliffs of Dover on Mars, no vast plateaus of limestone, no chalk downlands. This doesn't necessarily preclude liquid water on Mars; as we have seen, limestone may never have formed on a sulfur-rich Mars, But there are further problems with the warm, wet model. For a start, where did all the water needed to produce those outflow channels come from? If the biggest of them really were carved by gushing torrents of H_2O, how were the underground reservoirs that supplied this water replenished? It has been calculated that replenishment would need to have occurred on timescales of a few tens of thousands of years. None of the conventional explanations, maintains Hoffman, will do. It is possible, he admits, that there was a continuous source of volcanic water (he does not deny the existence of the huge underground reservoir of water *ice*), but if that were so then one would expect the outwash channels to be intimately associated with volcanoes. That is not the case. There are channels near the slopes of the great Tharsis mountains, but they are fewer in number than in many other locations. Hoffman admits that many of the Noachian distributory valley networks were carved by water, following localized melting caused by meteorite impacts. The later floods were not carved by water.

So if the channels were not carved by water, what made them? The answer, says Hoffman, is carbon dioxide – a mixture of gas and perhaps liquid, at high pressures, mixed with rock. 'Nothing based on water can flow at Martian temperatures,' he maintains, 'so the culprit must be avalanches of gaseous carbon dioxide and rocky debris'.

It is simply too cold, and always has been, for water to flow on Mars for any length of time. Hoffman's best guess, at the time of writing, is that the evidence from the Opportunity landing site suggests a few thousand years of cold, acidic shallow water at most. Carbon dioxide, however, is ubiquitous on Mars and always has been. And it could provide a powerful force to carve the Martian landscape:

Liquid CO_2 would have been ubiquitously stable in equatorial regions of Mars about 2 billion years ago, at burial depths of 50 m [160 ft] plus. CO_2 can also exist as a stable solid at depth, especially in the past when surface temperatures were lower. When a cliff collapse occurs this solid CO_2 depressurizes and – partially – melts. Thus liquids can be formed spontaneously from solid material, merely because it collapses. The liquid CO_2 then floods out through the dusty soil and literally explodes into Mars's thin atmosphere

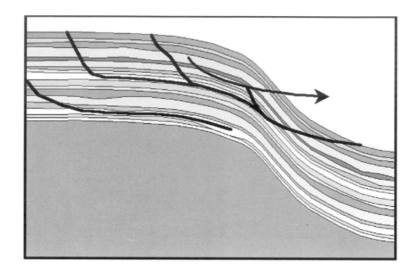

Nick Hoffman's White Mars hypothesis states that large quantities of carbon dioxide liquid and gas, released under pressure, are capable of shifting huge volumes of rock and soil and carving the wide, flat-bottomed outburst and outwash 'flood' channels such as seen all over Mars.

These diagrams show a sequence of events, starting with the collapse of a layered outcrop, the blocks of rock and soil lubricated by liquid CO_2. The second picture shows flows of rock and soil driven by carbon dioxide gas. As the flow thins out, it would turn into a dense, gas-supported avalanche, scouring the rocks beneath. The last pictures show both the density flow in motion and the resulting landscape – a terrain of isolated blocks, pyramids and mesas.

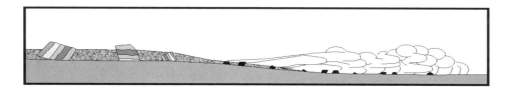

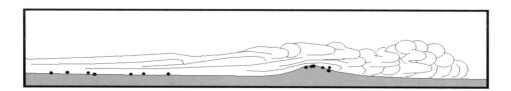

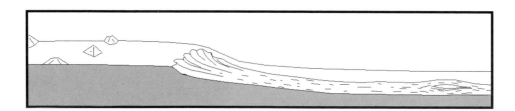

Pyroclastic flows are one of the most destructive aspects of volcanic eruptions and provide graphic evidence that rocks can be transported on a cushion of gas. This picture shows the destructive effects of the pyroclastic flow and surge that swept down Mount St.Helens in 1980.

with an internally generated pressure of at least 5 bars. The effect is like letting off a fire extinguisher into a bag of flour. The fine dust and sand of Mars' regolith is blasted into a huge dust cloud which roars down the valley like a fluffy avalanche (something called a cryoclastic flow).

However, this 'fluff' is powerful stuff. Like volcanic ash clouds on Earth, it can carry so much load in suspension that it flows like a liquid downhill and erodes and destroys anything in its way, carving the outburst 'flood' channels. Even solid rocky layers are blasted into fragments by the explosion and carried downstream as boulders. In this way the impressive boulders at the Mars Pathfinder landing site were transported and deposited by clouds of gas and dust, not by water.

In fact, density flows like this are common on Earth, and can be responsible for some quite spectacular landforms, Hoffman says. It is not just the infamous pyroclastic flows from volcanoes (like the one which did for Pompeii), but also snow avalanches and submarine turbidites (massive rivers of rock and cold water which can scour the sea bed for hundreds of kilometers and transport billions of tonnes of material). In the right circumstances, Martian CO_2-driven flows can exceed 500 kph (300 mph) and, most importantly, they need only a very low gradient to proceed – just 1° is enough. This is important because, some spectacular topography aside, the Martian surface is mostly flat.

But are the density flows Hoffman is talking about sufficiently powerful to explain the features seen on Mars? Remember, there are channels hundreds of kilometers long, and plains strewn with tens of thousands of boulders which much weigh several tonnes each. Can pockets of released gas mixed with rock and sand really do all this? Hoffman says they can. And he points to places on Earth where this has happened. The most spectacular examples are created by volcanoes. The 1996

'Without liquid water there cannot be life'

Nick Hoffman

eruption on the Caribbean island of Montserrat transported boulders weighing several hundred tonnes several thousand meters. They were carried not by lava, but by a mixture of thousand-degree gas, dust, and rock.

In January 2003, Hoffman made the headlines in the science media by claiming not only that the gullies discovered by Malin and Edgett had been formed by carbon dioxide gas, but that he had actually spotted them in the process of formation. Hoffman was always skeptical of the preferred explanation that the gullies had been formed by water. One possibility is that warm groundwater is being released to the surface through lines of springs or aquifers, high on the crater walls. But, Hoffman points out, if groundwater were responsible, you would expect the gullies to be seen in the warmer areas of Mars; in fact, they are nearly all close to the poles. In addition, if the gullies were effectively geothermal springs, you would expect them to be found near volcanoes. But in fact they are thousands of kilometers away. And finally, he says, the gullies begin at the tops of hills and sand dunes, rather than seeping out at the base like water should. 'The consequences of this for life on Mars are shattering. If similar mechanisms are responsible for all the recent gullies on Mars, then the near-surface life NASA is so desperately searching for may not exist', said Hoffman at the time. 'Without liquid water there cannot be life, and despite recent reports of more and more ice on the Red Planet, NASA has yet to find liquid water.'

Hoffman produced an annotated picture based on a Mars Global Surveyor image showing what looks like a layer of white snow or ice packing the 'alcoves' first spotted by Malin and Edgett. He explained that it

The key to Nick Hoffman's White Mars hypothesis is the ability of gas, rather than liquid, to move large amounts of solid material across a planet's surface. On Earth, pyroclastic flows from volcanoes can certainly move huge boulders, as can be seen here at Martinique.

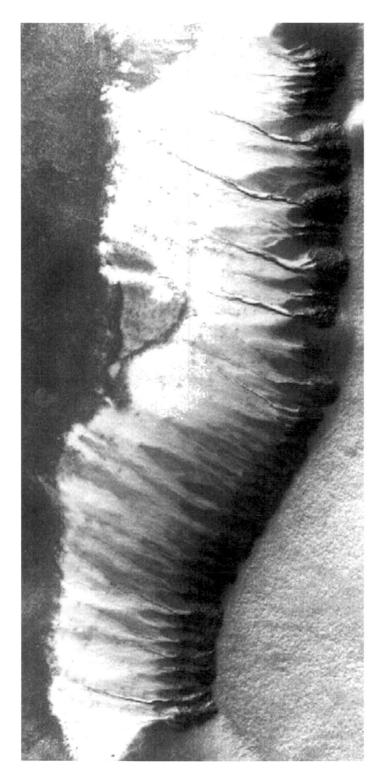

shows that in this area at least the flows are not merely 'recent' but are annual thaw features as the seasonal polar cap retreats. Because we can measure the ground temperature at the time, and can see extensive deposits of CO_2 snow and ice surrounding the flows, we know that they must be occurring at $-125°C$ [$-193°F$], far too cold for any aqueous (i.e. water-based) flow. As a consequence, thawing carbon dioxide must be invoked as a gaseous lubricant for the flows. But carbon dioxide doesn't melt on Mars; it sublimes directly from the solid. Instead of a trickle or gush of liquid pouring down the gully, the flow appears to be a flurry of boiling dry ice avalanching down the gully. This turbulent gas–solid mixture acts like a army of miniature hovercraft, carrying a shower of sand, dust, and tumbling rocks down the slope, carving out the gullies as it goes.

Hoffman's key argument relies on the temperature of the gullies. The most saturated brine possible will flow at around $-25°C$ ($-13°F$). But the gullies are far, far colder than this. 'What is indisputable is that the mean temperature at the time the image was taken was $-95°C$ [$-139°F$]', Hoffman says.

This is colder than the record coldest temperature on Earth, atop the Antarctic ice cap. The gullies themselves are even colder than this, and appear to contain extensive deposits of dry ice. Nothing based on water can flow at these tem-peratures. Battery acid [which would stay liquid far below the freezing point of pure water] freezes like concrete. And yet we see flows that have occurred under these conditions.

One possibility, suggested by Michael Carr soon after the gullies were discovered, is liquid carbon dioxide. But the trouble is that $-95°C$ is too cold even for liquid CO_2 – and the pressure is insufficient. Carbon dioxide sublimes at pressure

Nick Hoffman's White Mars model has been under concerted attack since he first postulated that carbon dioxide, rather than water, carved the channels of Mars. In January 2003, Hoffman claimed to have evidence – shown here – of active gully formation, where thawing CO_2 is eroding the rocks and regolith, which are tumbling down the crater walls before our very eyes. These gullies are located a few hundred kilometers north of the south Polar cap.

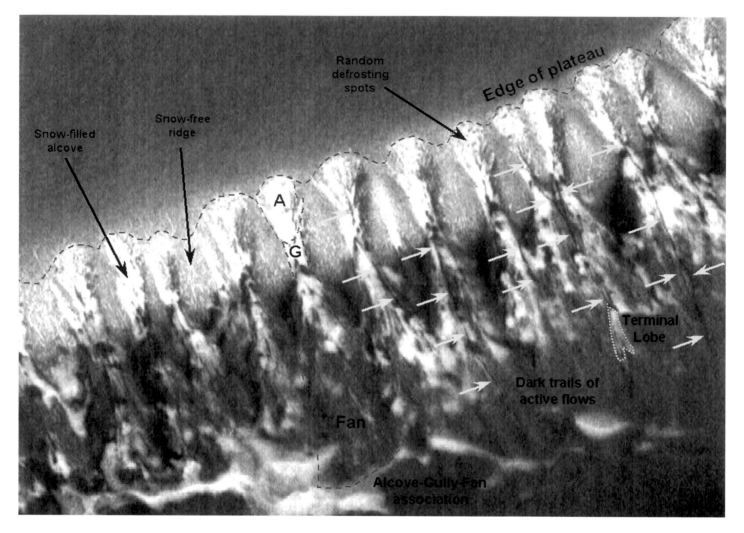

This annotated photograph, produced by Nick Hoffman, explains a possible mechanism of formation for the gullies (G) and alcoves (A). 'In this image,' he explains, 'lighting is from the south (bottom of image) so the dark spots and streaks within the channels cannot be caused by slope effects. They must be defrosting patches. Some of these random spots are noted. What is more important, is the thin dark linear features that wander down many of the channels from alcove to fan (shown in red). These streaks are the thin channels of active flows (depicted in yellow), in my interpretation.'

on Earth. Sunlight passes right through the dry ice, and is absorbed by 'dirty' snow and gravel at the base, which then warms up (as darker material absorbs more solar radiation).

According to Hoffman's theory, the thaw produces gaseous CO_2 under pressure, beneath the snow. This escapes forcibly by any means it can. Because of the steep slopes (typically 30° or steeper), the pressure destabilizes the ice and an avalanche is triggered:

Under the right circumstances, the avalanche involves some CO_2 snow, and some of the rocks and dust that were warmed in the sunlight. As they tumble down the hill together, the components are forcibly mixed, and warm rock touches cold

less than around 5 bar. What is happening instead is that carbon dioxide snow is evaporating in the spring sunshine from the bottom up, just as happens to melting snowpacks

dry ice. The dry ice boils away, generating more and more gas that supports the flow as a gas-lubricated avalanche, or a small density flow. While this contains no liquid at all, nonetheless it travels down the channel, transporting debris.

Hoffman's Mars is, to some scientists, a cold and unexciting place. He believes that Mars is frigid and dry today and has possibly always been so. Lakes and rivers of liquid water have never marked its surface, and there never was a huge ocean, bigger than the North Atlantic, swallowing the northern plains. He sees parallels between today's 'warm, wet Mars' theories (and the idea that the gullies seen by MGS are being carved by extant running water) and the 'optimistic' ideas about Mars that were being espoused in the late nineteenth and early twentieth centuries. He writes:

If Mars had water flowing in the past, then life could have, indeed should have, evolved. The present cold dry state of Mars must be some sort of long dormancy during an aeons-long planetary 'winter'. When springtime comes again, perhaps Mars will bloom and water will flow again?

The reason for the belief in a watery Mars, says Hoffman, is basically emotional:

There is a lot of funding in wet or blue Mars. It is part of the human psyche, isn't it, to want to not be alone. You can make Mars warm and wet if you really want to, but you have to really bend the model to do so. It is easy to forget how cold Mars really is. If you take off your blue goggles you will realize that there never was a warm and wet Mars. It is as much a figment of our imagination as Lowell's canals were. Mars is a heavily cratered remnant from the early solar system. We want Mars to be inter-esting – look at Mercury. We sent one probe there, took a look, saw loads of craters and thought, nah, it's boring. And we have never been back. I believe that if I had a time machine and traveled back to Mars 4 billion years ago, I would see a shining white planet – its surface covered with water and carbon dioxide ices.

One of the most scientifically literate works of sci-fi dealing with Mars is the *Red Mars*, *Green Mars*, and *Blue Mars*

trilogy by Kim Stanley Robinson. In three tomes, the author imagines a millennium or more of human colonization, during which the Red Planet is slowly engineered into a rough approximation of Earth. There are no little green men, no princesses, and no canals in these books – but according to Hoffman they are every bit as fantastical as the Mars of Ray Bradbury:

I am a big fan of these books – reading them was what got me interested in Mars in the first place. But the Mars of these books is no closer to reality than the wonderful Lowellian Mars of Ray Bradbury, which seems quaint, old-fashioned, and scientifically ludicrous to us now. But I believe that as we learn more about Mars we will come to see our current beliefs about water and a golden age in the past with oceans and great rivers, and probably life, as just as quaint and old-fashioned. Percival Lowell was no charlatan – he had doubts towards the end of his life. Unlike others, who are simply trying to make money from their ideas, I believe Lowell actually believed in what he said. He was no cynic.

Hoffman is not a lone voice. Although his views are fiercely contested by the warm-and-wet mainstream, there are other dissenting voices that question the possibility of an Earthlike Mars. Coming at the debate from another angle is Conway Leovy, a meteorologist from Washington University in Seattle. He too has his doubts that Mars was even more Earthlike than it is today:

We all like the more interesting paradigm, the one which says there was life and water and so on. Well, I have my doubts. I was once asked, Con, why are you so pessimistic? Pessimistic – it's an odd choice of word.

Leovy believes that the ghost of Percival Lowell still stalks Mars today. Where once we saw artificial rivers, we now see gigantic natural rivers, like the 8,000 km (5,000-mile) 'Martian Nile' that appears to pour out from the Argyre basin into the northern lowlands. The consensus – still – is that this feature was carved by running water. Agreeing with Hoffman, Leovy says that the problem is that you simply can't get Mars warm enough for long enough to allow this to happen.

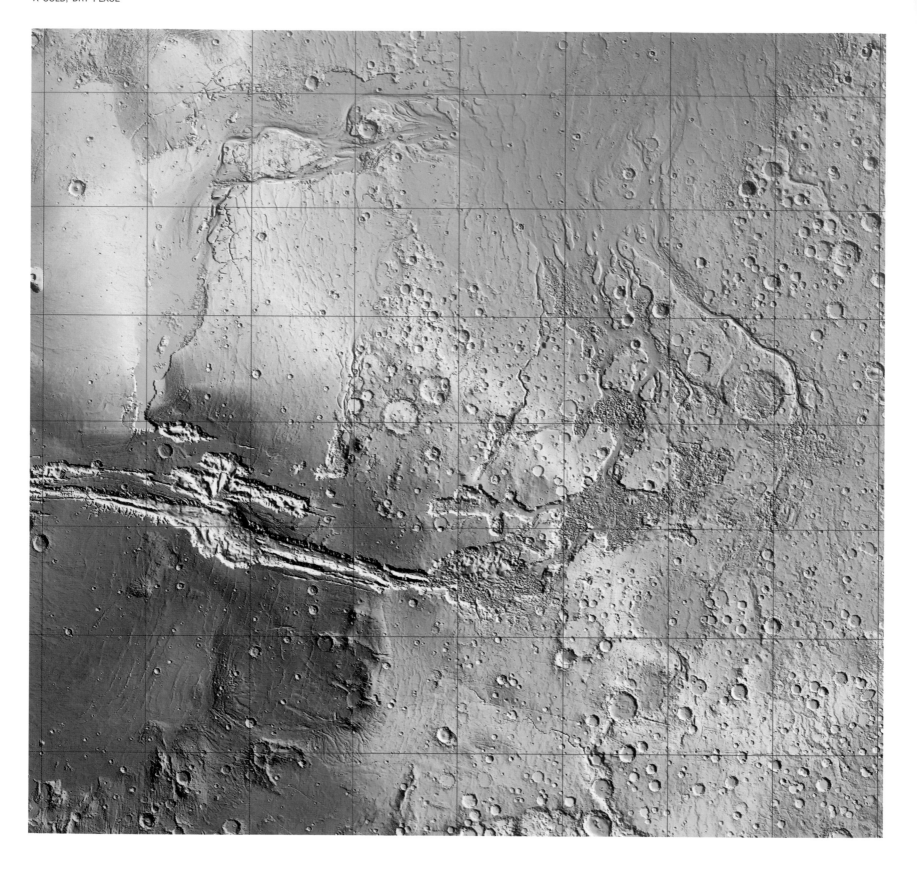

So how could these channels have been carved, if not by running water? Nick Hoffman's answer is pressurized carbon dioxide. Leovy proposes a rather less exotic solution: simple wind erosion will do the trick, he claims. 'We know that wind is actively eroding the surface and has been doing so for billions of years. We know that lots of areas are being exhumed by wind, and that there is deposition in others.'

Leovy's mechanism for the 'fluvial' channels starts with something opening up a fissure. This could be, he suggests, a lava flow eating into the soil. He points out that radar imagery of the Venusian surface from the Magellan space probe shows hundreds of channels, complete with meanders, braiding, and even oxbow lakes. 'If you knew nothing about Venus you would conclude that you were looking at water-carved features, yet no one thinks that these were indeed made by water', he says. This is because, at 450°C (850°F), the surface of Venus is even less conducive to liquid H_2O than is Mars. Instead, it is assumed that these channels were carved by lava flows – huge outpourings of low-viscosity magma gushing across the planet's surface.

Now, we know that Mars once had active volcanoes, and there is ample evidence for massive outpourings of lava over its surface too. Leovy imagines a scenario in which a lava flow carves out a gully in the surface and then cools. Then time gets to work. It is often said that with no rain and no rivers, erosion proceeds very slowly on Mars. That is true. But erosion on the Red Planet has time on its side. With a surface having lain undisturbed by the upheavals, swallowings, and collisions of plate tectonics, the wind – even the feeble Martian wind – has had aeons to get to work. It is wind that has carved the channels into the forms we see today, says Leovy. Even the most 'fluvial'-looking features, apparent river deltas in the floors of craters, for example, can be explained by debris flows.

(*Opposite*) Some of the most impressive valleys and channels seen on the Martian surface seem to emanate from the giant series of canyons in the Valles Marineris. This MOLA image shows a series of huge outflow channels running north from the Argyre impact basin that formed during the Noachian period of Martian geological history. Towards the top of the image the blue color indicates low elevation – and the final destination for any water that may have flowed down these channels.

(*Above*) The surface of Venus is far too hot for liquid water to flow, yet its surface is marked by dozens of long, ribbon-like channels. What carved them? Winds at the surface are so sluggish that it is hard to conceive that they were responsible. More likely is that the Venusian channels are carved by molten rock or exotic liquids exuded by still-active volcanoes. This goes to show that features on other planets may look familiar yet may be formed by wholly alien processes.

Phil Christensen suggests that the warm-and-wet hypothesis has been overstated too, although he does believe that many of the Martian channels were carved by water or ice. While he doubts that carbon dioxide can explain the apparently fluvial features on Mars (he believes that the gullies are essentially glacial features), he does think that NASA has made a mistake in pinning its hopes on finding water. He had this to say in 2003:

I think follow-the-water is a mistake. It is a double-edged sword. The danger is that it will come back and hurt us. There is a slippery slope. If you talk about water and life the public gets the unfortunate idea that Mars with no life is boring. Why can't Mars be interesting the way it is? I want to study Mars the way it is. Surely it is fascinating if it is different to Earth.

A further blow to the warm-and-wet hypothesis came earlier in 2003, when another paper indicated that large deposits of the mineral olivine had been found on some ancient areas of the Martian surface. Olivine – magnesium-iron silicate — is a pretty green crystalline substance that can be carved for jewelry (when it is known as peridot). It forms deep in the crust, in igneous rocks. As far as Mars is concerned, the presence of olivine is important because it is highly reactive with water – any prolonged exposure and it soon degrades into hematite (iron oxide) and other minerals. Finding large amounts of olivine which have been exposed for 3 billion years or more is, again, circumstantial evidence that large standing bodies of water never existed on Mars.

In 1965, the exploration of the planet was almost put on hold because of the 'disappointing' findings from Mariner 4. When later probes saw channels and gullies, planetary scientists were excited: Mars was coming back to life, if not literally then at least geologically. Vic Baker argues that some of the features

(Opposite) Some of the strangest landforms on Mars are to be found in the south polar region. Bizarre structures that look like giant insects or starfish have been dubbed 'spiders' and 'dalmatian spots'. Until recently their origin was a mystery, but a new theory suggests they are formed by little-understood processes going on under the frozen carbon dioxide of the ice cap.

seen on Mars look so much like they were carved by water that it is almost perverse to argue otherwise, even though it is theoretically possible that they were made by some exotic gas flow which we do not understand. White Mars has few fans.

THE SPIDERS OF MARS: CARBON DIOXIDE AT WORK

A lot of Mars looks a bit like the Moon. Bits of Mars look a lot like the Earth. But there are some places on Mars which could only be on Mars, truly alien landscapes that until recently have defied explanation. A good place to look for Martian oddities is on and around the polar caps. Nowhere on the planet is as strange or otherworldly as the icescapes around the poles, terrain shaped by the exotic behavior of not just water ice but also of solid and gaseous carbon dioxide – a gas which plays no role in the formation of landscapes on Earth, and whose role on Mars is so controversial. The cameras aboard Odyssey and Global Surveyor have found a bizarre menagerie of what are now known as 'cryptic material', which have been given names such as Dalmatian spots, oriented fans, fried eggs, and black spiders. Explaining these features has been a challenge.

The spiders consist of dark, circular patterns on the south polar ice. From some angles and under some lighting conditions they look a little like pine trees or shrubs (giving rise to the inevitable speculations from the face-on-Mars brigade), but exactly what they are has until now remained a mystery. Hugh Kieffer, of the US Geological Survey in Flagstaff, has come up with an ingenious mechanism for the spiders that shows just how un-Earthlike Mars is. Nothing like this happens on Earth – there isn't even a vague analogue to be found in any terrestrial environment.

Kieffer supposes that carbon dioxide is deposited at the poles either by direct condensation or by dry-ice snowfall into slabs of ice, maybe a meter (3 feet) thick. Initially the solid dry ice is full of dust, but as dust particles heat up in the sunlight they vaporize the CO_2 in their immediately vicinity and start to burrow their way through the slab. The carbon dioxide gas travels through the slab, accompanying the dust mote, eventually breaking through the bottom. The net effect of millions of such tunnelings is to create a cushion of gas

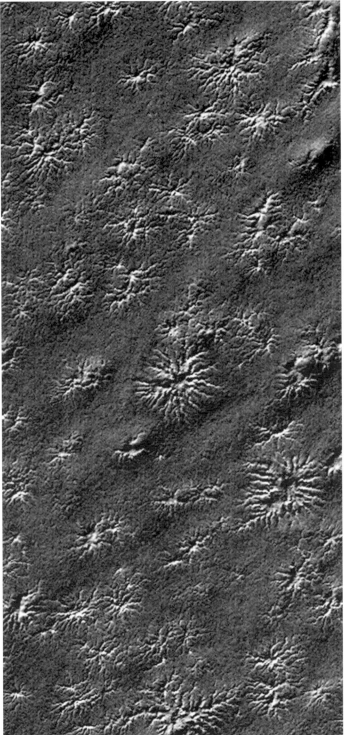

under a transparent layer of dry ice. This gas is under pressure – it has to support the weight of the solid gas above it – and it will escape by finding cracks in the dry ice. When a vent forms, the gas rushing to escape will scour the surface under the ice, creating linear marks and channels of dust. The 'spiders' we see from orbit are simply these scour marks seen through the transparent ice – and the 'Dalmatian spots' are probably the carbon dioxide 'geyser' vents.

Although Hoffman's Mars is out on a limb, most other planetary scientists are reluctant to dismiss it out of hand. 'You have to give him credit, no one has managed to shoot him down entirely', says Michael Carr. Perhaps his most convincing arguments revolve around the analogous formations he says exist on Earth – pyroclastic flows, undersea turbidites,

and so on. We know that gas can move boulders. Whether or not this has happened on Mars is unlikely to be decided for certain any time soon.

The latest results from the latest Mars probes, particularly the rovers Spirit and Opportunity, seem to quite literally pour water on Nick Hoffman's White Mars. There is strong evidence that at least parts of Mars were indeed fairly warm and fairly wet during at least part of their history. White Mars isn't dead yet – Hoffman thinks that carbon dioxide may yet turn out to be the dominant agent in the creation of much of the planet's surface – but the debate is raging more fiercely than ever. To settle it once and for all, it is becoming clear that we may have to go to Mars in person. But, sadly, that is proving to be a bigger job than we once thought.

Chapter 5
Going to Mars

'The logistic requirements for a large, elaborate mission to Mars are no greater than those for a minor military operation extending over a limited theatre of war'

Wernher von Braun (*The Mars Project*, 1952)

Percival Lowell imagined a lot of fanciful things in his life, but one notion he did not entertain too seriously was that one day humans would travel to Mars. He even wrote a (rather terrible) poem in which he yearned that 'Against hope hoping that Mankind may, In time invent some impossible way' to visit 'that other island across the blue'. He would have been flabbergasted to think that, just thirty-two years after his death, someone would be sketching out plausible plans for a human expedition to 'his' planet, and that just twelve years after that, robotic probes would actually be aimed at Mars.

We have been sending robots to Mars for four decades, but when are we going to see a man or woman on Mars? For the latter half of the twentieth century the answer was fairly consistent, and clear: 'in about twenty years' time'. Sometimes the timescale was shorter – during the heady days of Apollo, when just about anything seemed possible, a Mars landing was thought to be little more than a decade away. But if, in 1965, say, you had asked anyone seriously involved with the American space program when they thought they would see someone bounce down the steps of a rocket ship onto the Red Planet, even the most cautious would not have given a date much beyond 1985. If you had told them that, come the end of the twentieth century, not only had no one been to Mars, but also that the Moon had been abandoned for twenty-five years, they would have been astonished. What could have happened?

The reasons we have not been to Mars, or even back to the Moon, are of course far more complex. They include the Vietnam War, certainly; the death of Kennedy, possibly; and the end of the Cold War, most definitely. Add the horrendous

expense of exploring space, and a public reluctance, in America at least, to engage with a space program that is now more about scientific endeavor than national pride, and you have a recipe for what many observers maintain is a manned space program that is going nowhere.

Today, there is absolutely no chance, whatever some of the more enthusiastic commentators insist, of a crewed Mars mission being mounted before the end of the next decade. Despite the announcements by the US Administration in the spring of 2004, the reality is that a manned voyage will take at least two decades – probably three – to plan and execute, and that is if everything falls into place and the funds are available. Rules on safety and financial accountability are much stricter than in, say, the days of Apollo. The mission would almost certainly have to be an international effort, and, as the experience of the International Space Station has shown, large multinational space projects are good for the morale but extremely costly and inefficient ways of getting things done. It is quite possible that we will not see a Mars mission launched this side of 2050, and possibly even later. Maybe this forecast is unduly pessimistic; after all, few in 1930, say, would have predicted a Moon landing within forty years. And China may yet have plans for space exploration that will surprise us all.

At present, crewed spaceflight now consists of the Space Shuttle (suspended at the time of writing after the *Columbia* disaster), the International Space Station, and the Chinese and Russian space capsules. All have been criticized for doing little more than suck up huge quantities of cash that might be more profitably be spent elsewhere. Neither the Shuttle nor the Space Station are going anywhere. Both are

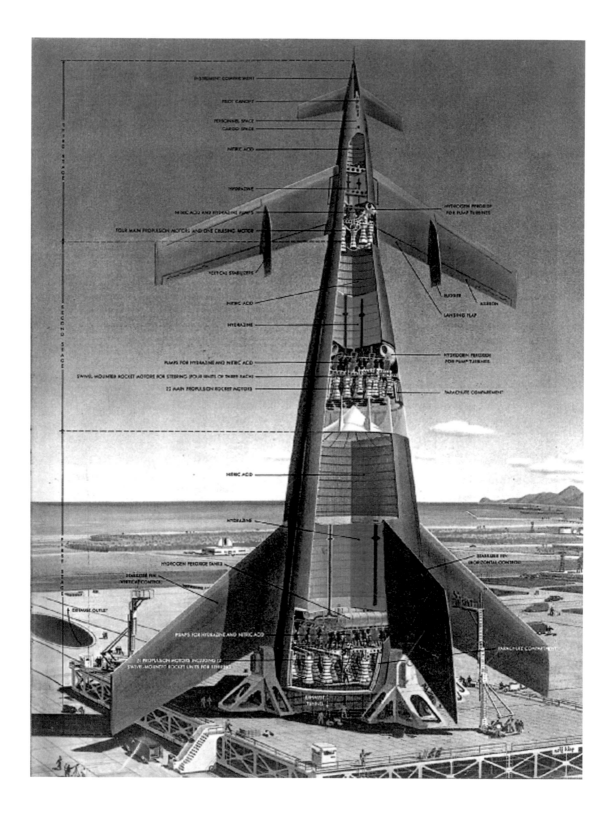

stuck in low Earth orbit, a 'place' that has been explored for more than forty years now. Since Gene Cernan, Harrison Schmitt and Ronald Evans returned to Earth from the Moon in 1972, no human has traveled more than about 600 km (375 miles) from Earth – a sobering thought.

It is all so very different from the heady days of the 1960s, when humankind was reaching for the stars. It is incredible to think that, a generation ago now, America's finest space engineers were talking seriously of building immense, 4,000-ton nuclear-powered rocket ships that would fly fifty-strong crews to the Red Planet at 100,000 kph (60,000 mph). They envisaged huge winged behemoths that could skate down onto the Martian ice caps, spinning spacecraft that could generate artificial gravity. It is strangely nostalgic to read of these early Mars mission designs, the brainchildren of Dr von Braun and his rocketeers.

Maybe we could have got to Mars by 1985; maybe not. It is possible that NASA (for it surely would have been NASA) would have run into unforeseen problems. In retrospect, it can be seen as lucky that no American astronauts were lost in space on any of the Apollo missions. Mounting a Mars mission with 1960s or 1970s technology would certainly have been far more risky than with twenty-first century machines. There would probably have been deaths, injuries certainly. We will never know, because it was never tried. But they came so close – money, real money, was spent. Studies were commissioned, contractors hired, mock-ups built. Humanity was heading for Mars and beyond, carried aloft in giant rockets the size of skyscrapers and powered by atomic furnaces.

As with so much about Mars, there was always the hint of science fiction about those early dreams and visions. And science fiction is what it turned out to be. When Stanley Kubrick's *2001: A Space Odyssey* was released in 1968, and showing an orbital station, a large base on the Moon, and

a nuclear-powered ship heading for Jupiter, few would have doubted that the first two would have been a reality by the turn of the century. What follows is a blueprint for a future that never was.

DAS MARSPROJEKT

The first serious attempt to design a manned mission to Mars was made by Dr Wernher von Braun. The former Nazi rocket scientist, who brought terror to London with his V-2s, and later redeemed himself by taking Aldrin and Armstrong to the Moon, dreamed of Mars even when Apollo was only a remote dream. As early as 1947, when he was still effectively a prisoner of the US Government, having been interned after his capture by the British and then handed over to their allies, he put his thoughts on paper in what was apparently an execrable novel (the book never saw print) about a crewed mission to Mars.

Although von Braun was no dramatist and his prose was leaden, he did know his rocketry, and the appendix to the book, a collection of mathematical proofs supporting his spacecraft design and the mission trajectory, was published as *Das Marsprojekt* in 1952, and translated into English a year later as *The Mars Project*. It was a far more riveting read. By now, the exuberant von Braun and his German colleagues were free men, and US citizens to boot.

So what was von Braun's plan for Mars? It couldn't have been more ambitious. Remember, he wrote this when no one had been into space – indeed, no*thing* had been into space. There were no computers, no telemetry guidance systems, and no communications satellites. Humanity's best photographs of Mars showed an orange and greenish globe with white bits at the poles. American engineers were tinkering with captured V-2s (and Russian engineers were doing the same thing), but it was to be another five years before Sputnik put a rocket up America and sparked the space race. Von Braun planned to send ten 4,000-tonne ships straight to Mars. That is 40,000 tons of space hardware, each crewed by seventy people. To this day, the largest rocket ever built, the Saturn V launch vehicle used to send the Apollo astronauts to the Moon, weighed just over 2,700 tonnes.

Von Braun thought it should be possible to get straight

(Opposite) Just imagine. This artist's impression taken from *Collier's* magazine from the 1950s shows the sort of giant Mars ship that might have been built had Wernher von Braun had his way. His vision of an assault on Mars makes fascinating reading today; in the pre-computer age, his astronauts would have performed their navigational calculations using slide rule, sextant, and pen, and communication with Earth would have been almost impossible.

(*Above*) Rocket man. Wernher von Braun was an extraordinary figure. After he was captured by the American forces in Germany, he was brought over to the United States to help develop the nascent rocket program. Once he was granted full US citizenship, he took to American life with gusto, working all hours of the day and going for marathon swims in the Pacific Ocean at night.

(*Opposite*) Detail from the Chesley Bonestell painting illustrating the 1956 book *The Exploration of Mars*, by Willy Ley and Wernher von Braun. It shows a group of astronauts contemplating a rocky outcrop, perhaps searching for signs of fossil life.

to Mars after assembling his fleet in orbit, and he proposed 950 ferry flights to haul the Mars-ship parts into space, and assembly would take place *in situ* – no space station assembly base was assumed. Getting all this kit into orbit was going to take a lot of fuel. Von Braun calculated that he would need nearly 5.5 million tons of nitric acid and alcohol propellants, which is about ten supertanker loads. Attempting to play down the sheer scale of his proposal, he pointed out that this amount of fuel was only 'about 10 percent of an equivalent quantity of high-octane aviation gasoline that was burned during the six months' operation of the Berlin Airlift'. Ten Berlin Airlifts'-worth of fuel. Whichever way you looked at it, it was a tall order. And the cost? Well, von Braun estimated $2–3 billion dollars. Multiply that by six (the usual factor when it comes to initial estimates for gargantuan space projects) and then by a further 50 to allow for inflation, increases in fuel and labor costs, and

'Columbus knew far less about the far Atlantic than we do about the heavens'

Dr Wernher von Braun

so on, double it again for good measure and you get a present-day cost of getting on for $2 trillion, probably not far off the mark. This was some expedition, comparable in spirit and scale to the great exploration carried out in the fourteenth century by the Chinese admiral Xeng He, who ventured forth in liner-sized superjunks to investigate the planet, or to the European seafarers who charted the world in the fifteenth and sixteenth centuries. 'In 1492,' von Braun wrote,

Columbus knew far less about the far Atlantic than we do about the heavens, yet he chose not to sail with a flotilla of less than three ships, and history tends to prove that he might never have returned to Spanish shores with his report of discoveries had he entrusted his fate to a single bottom.

Seven of von Braun's ships were huge, interplanetary craft not designed to land. Huge assemblies of girders and pressurized, thin-skinned spheres (nothing then was known about the potential hazards from radiation in space or from micrometeoroids) and girders, they resembled space stations broken free from Earth orbit. Three ships would actually land on Mars. One, a 125-tonner, would glide down to the ice on one of (presumed perfectly smooth) the polar caps, landing on skis. This would have to be a giant craft, probably bigger than a Boeing 747, with huge delta wings. It would be an impressive sight indeed.

It was not known at this time quite how thin the atmosphere of Mars is, and how impractical such a vehicle would be (you *can* glide through the Martian atmosphere, but you would need wings the size of football pitches to do so). Von Braun assumed a surface atmospheric pressure of about 85 millibars (the prevailing best-guess estimate at the time). This is enough for liquid water to be stable up to 44°C (111°F). He also assumed that the mean yearly Martian temperature was 9°C (48°F), but that 'these figures require further confirmation'. Why the polar caps? Well, von Braun reasoned that these are

When Wernher von Braun sketched out his plans for a manned Mars mission, he envisaged a fleet of gigantic spaceliners winging their way from Earth to Mars, and crewed vessels putting down on one of the polar caps, on skis. This is Chesley Bonestell's depiction of the dramatic arrival.

the only places on Mars guaranteed to be smooth enough for a landing:

Considering the risk attending a wheel landing on completely strange territory at relatively high speed, it is assumed that the first landing boat will make contact with the Martian surface on a snow-covered polar area, and on skis or runners, minimizing this risk.

The first men on Mars (Von Braun envisaged only all-male crews), probably three of them, would then undertake what would be the most heroic expedition in history: to the Martian equator, using powered rovers, and transporting all their supplies with them. When they got there they would somehow bulldoze flat a kilometer-long landing strip onto which the other two landers, wheeled this time, could put down safely. As soon as

The first Mars bases, if they are ever built, will probably resemble the scientific outposts we have constructed in Antarctica. Life will be hard in such a hostile environment, with freezing temperatures, solar radiation, and, of course, no air to breathe.

the three teams of astronauts met up, they would take the wings off the two wheeled landers and haul them upright, ready to blast off from the surface. Meanwhile, the remaining seven ships would remain in orbit, tended by a skeleton crew.

Now exploration of Mars would truly begin. At this time it was assumed that not only did Mars have a reasonably thick atmosphere, but that it was almost certainly home to life. Von Braun wrote that on their 400-day survey of the surface (plus, of course, the several weeks that the pole–equator trek would have taken), the astronauts would spend much time 'surveying the flora and fauna'. Quite what he expected them to find is anybody's guess, but in the 1950s the consensus was still that Mars would have, at the very least, lichens, and probably things akin to insects as well. Not to

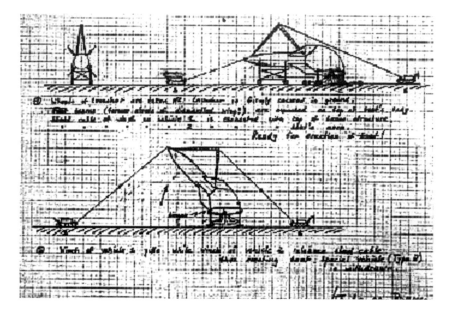

This sketch was made by Wernher von Braun and shows the erection of the Mars ascent ship on the surface. His astronauts would have needed to be skilled mechanics and physically fit and strong.

mention canal-builders. One main job would be to explore the canals, which were still confidently assumed to stride across the Martian surface.

One of von Braun's assumptions was that any Mars mission would require a large number of humans, to fly the spacecraft and tend to them, to scout for landing sites, and to carry out the exploration. When Ferdinand Magellan and Christopher Columbus set sail, not only did they have no idea what they would find when they arrived at their destination, but they knew that they would lose all contact with their home countries for months or years at a time. Communicating with astronauts on Mars was, in the early 1950s, thought to be equally impossible. Von Braun assumed that the best radio communication would consist of slow, dot-and-dash telegraphy. Any expedition would have to be entirely self-sufficient. If a spacecraft broke down, or a system failed, then the experts would need to be there, on hand, to fix it – hence the need for gargantuan ships and dozens of crew. In fact, space exploration has not been like that. The Apollo Moonships carried just three astronauts, and when a serious breakdown *did* occur, on Apollo 13, repairs were made by utilizing the help and expertise of

dozens of skilled engineers – who were back on Earth, at mission control. As space historian David Portree points out in his NASA monograph *Humans to Mars*,

An army of personnel, including scientists, formed part of each Apollo expedition, but remained behind on Earth. This separation was made possible by communication advances von Braun did not anticipate. In 1952, von Braun stated that television transmission between Earth and a Lunar expedition would be impractical. Sixteen years later, Neil Armstrong's first footsteps on the dusty, cratered Sea of Tranquility were televised live to 500 million people.

Von Braun's early Mars plan was ambitious and audacious. He saw humanity's destiny in space, and he was prepared to push the envelope to the limit to show that we could get there. This was 1952; even color TV was a novelty. Yet at least one man was thinking of going to Mars, and not only that, he had a plausible way of getting there. Von Braun himself cheerfully admitted that his Mars project was, in essence, no more than a back-of-the-envelope sketch. He acknowledged that there was no way to navigate his huge fleet of ships, that the technology to keep groups of men was not there, and that anyway in the late 1940s we knew very little about Mars at all. But it was typical of an age when technology was progressing ever faster that such a fantastical plan was thought of at all. And in fact, von Braun's Martian flotilla was by no means the most fantastical idea to emerge from the rocket scientists' dream-pads in the coming years.

Over the next thirty years or so, von Braun and his colleagues would refine their Martian dream. No more gargantuan fleets weighing tens of thousands of tons, but fast, streamlined missions relying on new, untried but promising technologies based on the power of the atom.

NUCLEAR TRAVEL

The main problem with getting anywhere in space is that it requires a colossal amount of energy to do so. Earth's gravity well is deep, and just getting onto orbit is a job in itself. Getting from there to Mars, and then back again, stretches the engineering parameters to the limit, even if you have unlimited funds at your disposal. Apart from a few experimental probes, such as the Deep Space mission that has trialed a small-scale version of an ion drive, all humankind's adventures into space so far have been courtesy of the chemical-fuel rocket.

Rockets, which have been around for over a thousand years, are, in principle simple machines. You burn a mixture of fuel and an oxidizing agent in a chamber. The resulting conflagration is then directed outwards, through some sort of nozzle. Newton's laws do the rest, and the nozzle, explosion chamber, and the bits and pieces attached to them are thrust smartly in the opposite direction. The biggest rocket ever flown, the Saturn V, was capable of lifting some 120 tonnes of payload into Earth orbit, and propelling 45 tonnes to the Moon. It was by far the heaviest successful flying machine ever made. Two-thirds of the launch weight was fuel – liquid oxygen and kerosene for the first stage, and oxygen and hydrogen for the second and third stages.

Most of this fuel is not needed to fly the Apollo to the Moon, or even to lift the Saturn launch vehicle itself off the ground. The depressing realities of the laws of physics mean that most of this fuel is needed to get the fuel itself moving – and back again, as the fuel needed to escape from the Moon must first be taken there. Now, going to Mars will be far more of a difficult proposition than getting to the Moon. Not only do you need an enormous rocket to escape from the Earth's gravity and accelerate the crew towards Mars (or a fleet of smaller rockets to assemble the ship in orbit), you also need to decelerate and fall into orbit when you get there, and descend to the surface. Then, which is the real challenge, you need to get the crew home again. All this means that you need to carry a lot of fuel to Mars with you (getting off the Martian surface into space is about twice as hard as getting off the Moon).

So, if you could find some way of getting there without burning kerosene or hydrogen, a Mars mission would become a lot more attainable. One way, mooted in the early days of Mars mission planning, was to use a nuclear rocket. A nuclear rocket would free you from the need to build an enormous fleet of 'Death Stars' (Mars visionary Bob Zubrin's dismissive nickname for grandiose ocean-liner Mars concepts) in Earth orbit, ready to make their stately way to Mars.

When the Saturn V rocket was signed off by von Braun and his engineers, it was becoming clear that it would represent the apogee of NASA's heavy-lift designs. There wasn't going to be a bigger, better rocket, at least not any time soon (although von Braun was nurturing plans for an even more gigantic machine, called the Nova). So any piloted Mars mission was going to have to rely on the Saturn to get itself into space. But for the flight to Mars, and from Mars back to Earth, another technology would be needed for a cheaper, more streamlined mission than von Braun's earliest plan. (In fact, von Braun himself conceded that his 'Death Star' approach would probably never be feasible, on cost grounds alone, but that it was 'exciting as well as instructive' to show that humans could reach Mars using available, 1950s technology.) He speculated at the time that within a decade or so some sort of nuclear propulsion system could be devised that would dramatically reduce the amount of fuel needed for a crewed Mars mission.

In fact, atomic missions to Mars were on the drawing board even before NASA was created in October 1958. In the previous year, the National Advisory Committee on Aeronautics (NACA, out of which NASA grew) commenced research into nuclear and electric propulsion for spacecraft, specifically aimed at interplanetary flight. This is impressive: just a dozen years after the end of World War II (and only a month after the launch of Sputnik 1), people were already thinking of the postrocket age. In fact, some even suggested that the space program should be run by the Atomic Energy Commission.

The work on the nuclear Mars rocket was done by engineers at the Lewis Research Center in Cleveland, Ohio. The first official NASA study of a Mars mission began in 1959 – before a single American had even flown in space – after the nascent space agency had been granted funding by Congress. The first NASA Mars mission envisaged a vastly scaled-down version of von Braun's plan. One Mars ship, either assembled on the ground and blasted into orbit atop a Saturn-sized rocket, or, more likely, assembled in space after a number of launches, carries a crew of seven men (the sex was specified) to Mars, accelerated into Martian transfer orbit by a high-thrust nuclear rocket. On arrival at Mars, two men make their way to the surface, spend some time exploring, then blast their way back into space using a conventional chemical rocket to rendezvous with their five colleagues in the 'mother ship'. This then accelerates back to Earth, where a landing vehicle separates and takes the crew home. The Lewis engineers were even so bold as to suggest a date for the first launch of a manned landing

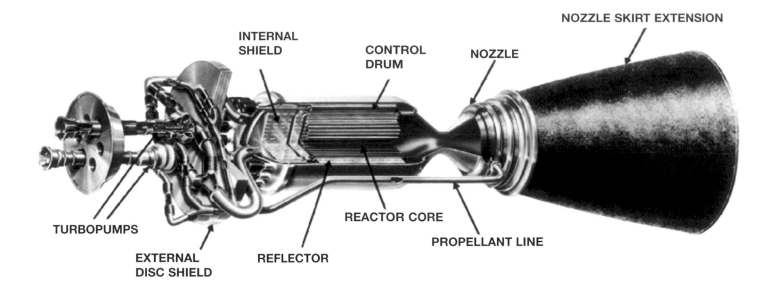

INTERNAL
SHIELD

CONTROL
DRUM

NOZZLE

NOZZLE SKIRT EXTENSION

TURBOPUMPS

REACTOR CORE

PROPELLANT LINE

EXTERNAL
DISC SHIELD

REFLECTOR

This cutaway drawing shows a NERVA engine, the nuclear rocket that America built in the 1960s and which was almost ready to fly in space before the program was cancelled.

mission to Mars – 19 May 1971 (this was the date of a particularly favorable opposition). The scientists cautioned, however, that 'this is not meant to imply that actual trips are contemplated for this period'.

How would a nuclear rocket work? The principle is quite simple, although not as simple as a conventional chemical rocket. A nuclear reactor, similar in design to the ones that generate electricity on Earth, would provide the heat source. Hydrogen gas would be passed over the fuel rods, causing it to heat up several thousand degrees until it became an ionized gas – a plasma. This would expand rapidly, and be ejected out of a nozzle, producing thrust. This doesn't seem too different from a chemical rocket, which again can use hydrogen as a fuel, but the key thing about such an atomic engine is that the hydrogen is not burnt, only heated and expanded. This means that no oxygen, or oxidizing agent, needs to be carried, reducing the volume of fuel by half. In addition, because of the extremely high temperatures that could be achieved by a nuclear reactor, a much greater thrust could be attained than with any chemical rocket. This is because thrust is proportional to the square of the velocity

of the material ejected from the nozzle: you double the speed to get four times the thrust. If an efficient nuclear rocket could be built, it would open the door to solar system exploration.

THE DEATH OF KENNEDY

When the nuclear bomb was invented, scientists quickly reasoned that the enormous power of the atom could be put to use in powering a spacecraft. Stanislaw Ulam, a Polish émigré rocket scientist and originator of perhaps the most outrageous space travel concept to date (see below), remembers that:

The idea of nuclear propulsion of space vehicles was born as nuclear energy became a reality. It was an obvious thought to try and use its more powerful concentrations of energy to propel vehicles with very large payloads for ambitious voyages of space exploration or even for excursions to the Moon.

In fact it was the great physicist and visionary Richard Feynman who was the first to talk about using an atomic reactor which would heat hydrogen, forcing it out of the back and propelling it forward – making a simple but effective and highly

efficient rocket. His ideas later became the basis for a number of projects to build nuclear-powered aircraft and spacecraft, including atomic-powered ballistic missiles, nuclear-powered warplanes, and the NERVA (Nuclear Energy for Rocket Vehicle Application) program. Plans for atomic planes were quickly shelved, but the NERVA program survived until 1973 – outlasting the Apollo lunar program. Project NERVA's remit was to develop ways of using nuclear reactors to power interplanetary spacecraft, and several of NASA's most successful spacecraft – such as the Voyager and Galileo probes to Jupiter and the outer solar system – have used small, plutonium-fuelled power packs (not reactors) to power their electrical instruments (actual propulsion, as opposed to electricity generation, was achieved using wholly conventional rockets on these probes).

Speed is one of the main reasons for wanting to develop atomic spaceflight. The need for a fast, powerful means of propulsion was accentuated by the discovery that space was not a harmless vacuum, where the only threats to crew safety would be psychological. Interplanetary space is full of potentially dangerous ionizing radiation, and once this was understood, Mars mission planners realized that the ships would have to be fast and the crew compartments would have to be heavy. Von Braun had envisaged thin-skinned passenger compartments for his Mars fleet: it was now clear that flimsy walls would not provide sufficient protection for the astronauts. Thick shielding – several centimeters of metal at least – would be needed to block radiation. For especially dangerous times, such as during a solar flare (when radiation streams off the Sun's surface), extra shielding would be required. Much of the subsequent work in Mars mission design has been about finding ingenious ways of protecting the crew. Some designs have incorporated drinking water tanks as radiation shields; a Russian plan favors using compressed waste matter to line the inside of the crew compartment. By cutting journey times, a nuclear rocket (which could reach Mars in a couple of months, say) would massively decrease the time of exposure and reduce the amount of shielding needed.

A test NERVA engine, mounted on its stand. The 'atomic rocket' sounds straight out of Flash Gordon, and perhaps few people realize how close it came to reality. There is now talk of reviving the nuclear propulsion program.

This test resulted in the rather catastrophic failure of a Kiwi engine. The resulting explosion was the worst-possible PR for the NERVA program.

Although the 1971 opposition of Mars had been proposed in the 1950s as a tentative date for the first Mars mission, no one seriously believed that America could go from no space travel to landing on Mars in eleven years. In 1960, NASA set out its plans for space exploration, based on what it believed was a realistic timetable. A space station was planned for the late 1960s, with a series of lunar exploratory missions to follow. The first lunar landing was to be some time in the mid-1970s. At the same time, work would begin on Mars. A Martian landing could take place shortly after the lunar missions. Mars and the Moon would be explored concurrently.

When, on 25 May 1961, President John F. Kennedy announced that an American would walk on the Moon before the decade was out, it was initially seen as a blow to hopes of getting a person to Mars. Kennedy's decree was a tall order, by anybody's standards. America had only just entered its space age, and the Soviets were clearly ahead; getting to the Moon by 1970 would mean that the entire thrust of the US space program would have to be focused on this one goal. Mars would have to wait. But although prioritizing Apollo meant canceling plans for the 1971 Mars trip, the advent of the Saturn V cheered the Mars enthusiasts. This was such a huge launch vehicle that it could easily lift an atomic-powered Mars rocket into orbit.

The NERVA project had actually begun in 1946, at the Los Alamos National Laboratory in New Mexico (it was at Los Alamos, the birthplace of the atomic bomb, where Feynman had had his idea). Initially, the idea of the project, which was run by the Atomic Energy Commission (AEC) and the US Air Force, was to see whether this technology had any military applications. One early idea was a nuclear-powered missile that exploded on landing. A ground-based test engine was built, named Kiwi because it was never going to get off the ground. The nuclear work being done in New Mexico led to pressure being applied (in the end unsuccessfully) on the government for the new space agency to come under AEC control. Foremost among those proposing this transfer was New Mexico Democrat senator Clinton Anderson. Even after NASA was made a civilian agency under federal control, Mr Anderson, in a classic case of pork-barrel politics, continued to support the nuclear rocket program throughout the 1960s, right up until his death.

The Kiwi atomic rocket was tested for the first time in July 1959 at the Nuclear Rocket Development Station at Jackass Flats in Nevada. The test was a success. A keen Senator Anderson arranged for delegates at the Democratic National Convention to be on hand to witness the second test, also successful, in July 1960. He even managed to get a commitment to nuclear rocketry built into the Democratic Party manifesto. Although the announcement of Apollo meant that a Mars landing was not going to happen as quickly as some had hoped, NASA was enthusiastic about developing a nuclear rocket. In July 1961, Aerojet-General Corporation won the contract to develop a 200,000 lb thrust NERVA flight engine. It was planned to lift an atomic rocket into orbit, aboard a Saturn V, in 1967. A new engine, Kiwi-B, was tested in

'We have a good many areas competing for our available space dollars, and we have to channel it into those programs which will bring a result'

President J.F. Kennedy

December 1961. The test was not a success. The reactor started shaking itself to bits, and blobs of molten uranium were ejected at high speed across the test site – hardly reassuring. Two further tests were just as unsuccessful. Clearly, a lot more work was going to be needed.

Kennedy was not impressed, saying that as getting to the Moon was a priority, the space test of the nuclear rocket would have to be postponed until additional tests had been completed and the reactor rocket could be seen to work on the ground. On 12 December 1962, the president said, 'We have a good many areas competing for our available space dollars, and we have to channel it into those programs which will bring a result – first, our Moon landing, and then consider Mars.' In the end, we never got to find out whether or not Kennedy would have approved the atomic rocket. The President was murdered on 23 November 1963, just weeks after the NERVA program seemed to have gained its second wind. The new president, Lyndon B. Johnson, cancelled all plans to take NERVA into space within a month of taking office, despite his close friendship with Clinton Anderson. Ground-based tests continued, but NERVA was finally killed off in 1973. So was it Kennedy's assassination that killed off the Mars program?

Probably not, it seems. Interest in nuclear propulsion was already waning. The idea of sending large fission reactors into space was increasingly seen as a less-than-ideal solution to the problem of getting humans to Mars. The growing environmental movement pointed out, not unreasonably, that should an accident occur, NASA would have a potential catastrophe on its hands. Opposition to nuclear power in space grew throughout the 1970s and after, to the point where even the small atomic 'batteries' attracted the wrath of the environmentalists. These Radioisotope Thermal Generators (RTGs), that generate electricity from the decay of non-weapons-grade Plutonium 238, were carried aboard a host of spacecraft, including the Pioneer probes to the outer solar system, the

Galileo mission to Jupiter, and the Cassini mission to Saturn. They have proved safe and dependable; notwithstanding this, the green movement almost managed to get the launch of Cassini stopped in 1997, and it may have been green activists who took potshots at the truck carrying Galileo across America from California to Florida. To oppose the environmentalists' claims, a Web-based organization, nuclearspace.com, gives an often entertaining and robust counterargument in support of nuclear power.

During the 1960s, the NERVA program progressed to the point of developing nuclear engines that were very nearly flight-ready. In its final form, the NERVA engine had met nearly all the stringent requirements imposed by NASA chiefs. In retrospect it is clear that the American space program very nearly acquired a powerful, lightweight, and high-tech replacement for the chemical rocket. Impressive as a Saturn V is, it is essentially no more sophisticated than a steam engine; an atomic rocket represents a technology shift. Fortunately, the expertise gained during the NERVA program was not lost. The project was effectively mothballed rather than obliterated, and it is certainly true that should the NERVA project ever be in some way revived, a workable nuclear rocket could be in NASA's hands in just a few years.

SITTING ON TOP OF A BOMB

Squirting plasma out of the back of a rocket was one way of utilizing the power of the atom to travel through space. But there is another, far more dramatic means of using nuclear energy to get off the ground. Don't bother with rockets at all: no reactor, no nozzles, no reaction chambers. Simply place an atom bomb (actually, a whole series of atom bombs) under your sturdy spaceship, stand (very) well back, and fire. It sounds crazy, but for a short time in the 1950s and early 1960s it looked like it might have well been tried. Project Orion was the name given to a secret American study into the feasibility of using what was in effect an atomic-bomb machine-gun to power giant spacecraft to Mars and beyond. Mad as it was – and undoubtedly dangerous – there is something quite wonderful about Orion, the story of which has been well told in a gripping book by George Dyson, the son of one of the project's founder members.

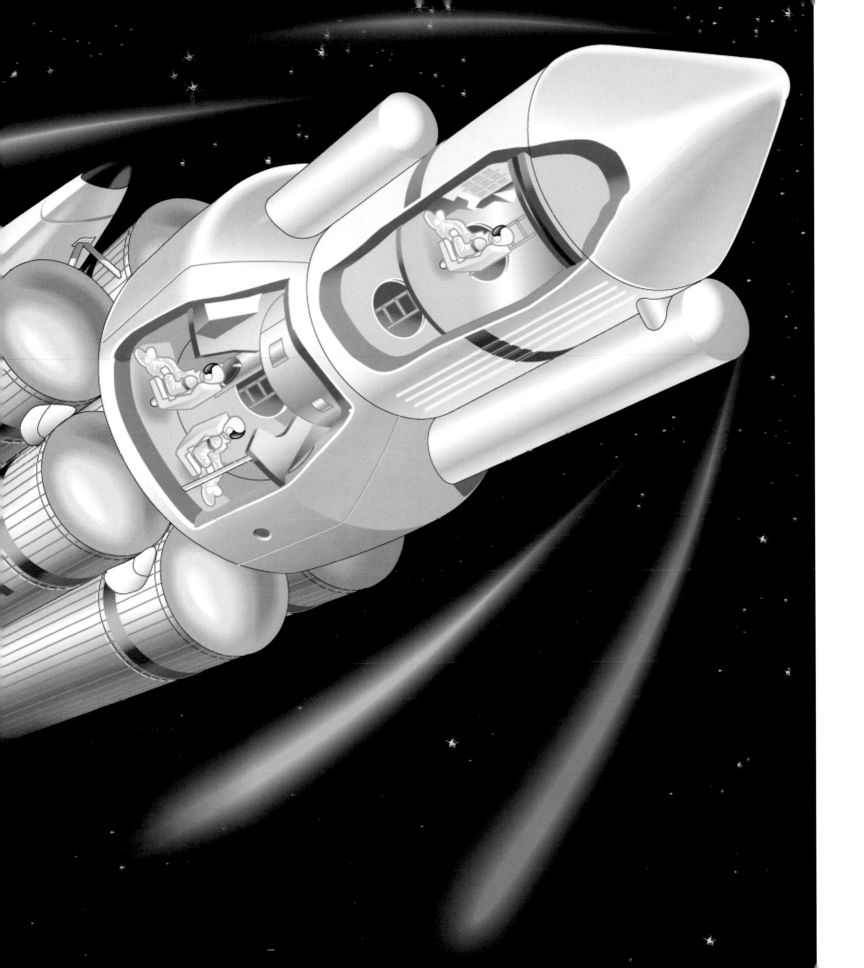

(Previous spread) The Orion Project was probably the most ambitious in NASA's history. The construction of a huge interplanetary – or even inter-stellar – vehicle would have given humankind the keys to the Cosmos. It never happened, of course, but what if it could be made to work ...

An Orion launch would have been the most spectacular pyrotechnic display in history. By comparison, the blast-off of a Space Shuttle or even a mighty Saturn V would have seemed like a penny firecracker. The spacecraft, 40 m (135 ft) wide and as high as a twenty-story office block, would climb into the sky powered by the fires of hell itself. Atomic explosions, 300 of them – one every half a second and each with a yield of between 0 and 20 kilotonnes – would push the 10,000-tonne ship into orbit. Each explosion would add about 30 kph (20 mph) to the monster-ship's speed. The noise would have been indescribable, and the blast-off would have been heard and seen across an entire continent. According to one of its creators, 'the first flight of that thing doing its full mission would be the most spectacular thing humans had ever seen.'

Attaining speeds no ordinary rocket could hope to achieve, this bizarre, madcap spaceship could get a large crew – and by large, the scientists imagined a whole spaceborne community of pilots, scientists, medical staff, and even cooks – to Mars in three months, or Saturn in three years. In the late 1950s, when the project was in full swing, with a team of 50 scientists and a $10million annual budget, the scientists declared 'Mars by 1965, Saturn by 1970!' They even discussed building giant, million-tonne Orions that could get to the nearest star, Proxima Centauri – over 4 light-years away. That the idea of a rocket powered by nuclear bombs was given the time of day by the usually cautious NASA and the US government was extraordinary enough. The fact that the research – much of it still classified – showed that the thing could actually have worked is quite startling.

The whole idea was dreamt up by Stanislaw Ulam, who had worked on the earliest nuclear weapons. Together with the British physicist Freeman Dyson and his American colleague Theodore Taylor, Ulam and the Orion team sketched out plans for a revolutionary spacecraft that the laws of common sense said could not fly. But fly it did. 'Those were extraordinary days', says George Dyson. 'They allowed these scientists to get their hands on 400 lb [180 kg] of C4 plastic explosive and play with it. That would never happen now.'

The go-ahead for a 'Feasibility Study of a Nuclear Bomb Propelled Space Vehicle' came in June 1958, and was given by the Department of Defense's Advanced Projects Research Agency (ARPA) – the same agency that gave birth to the Internet a little over a decade later. The ships were to be built by the General Atomic division of General Dynamics, which also constructed nuclear submarines. 'Preliminary studies have indicated that it is conceivable to use nuclear bombs as the energy source to propel a very large, manned vehicle to very high velocities', the contract read. The initial feasibility study was funded to the tune of a shade under a million dollars – about $16 million in today's money.

Ulam thought 'out of the envelope'. He was part of the Los Alamos team, and when he and the other scientists saw the world's first atomic explosion, at the Trinity test site in New Mexico in 1945, most thought that they had wrought the end of the world. But Ulam believed he had seen the key to the stars. As he saw the mushroom cloud rise over the desert, he imagined a spaceship rising atop it. Here was a power source far more potent than any rocket fuel. One way of harnessing an atomic bomb would be to explode it in a chamber and channel the force of the explosion in a particular direction. But Ulam realized that no material on Earth was strong enough to contain a blast that could flatten a city. Instead, he dispensed with a rocket chamber altogether. His idea was, in fact, one of the simplest spacecraft ever designed. The craft itself would be a squat, streamlined tapering cylinder, shaped not unlike a huge artillery shell. At its base would be a huge plate of steel or titanium, a couple of meters thick and coated with a thick layer of radiation-absorbing graphite.

Now comes the crazy bit. The Orion ship would have carried a supply of very small atomic fission bombs – mini-versions of the devices that destroyed Hiroshima and Nagasaki. Every half a second or so, one of these bombs would be shot out of the back of the spacecraft, through a hole in the plate, and detonated a few meters to the rear. The resulting explosions would hammer the ship upwards into the sky in a 5-minute cacophony. The propellant – as opposed to the nuclear fuel – could be any material that came to hand. In a chemical rocket, the fuel, heated by its own combustion, is the pro-

'The problem of course is to convince oneself one can sit on top of a bomb and not be fried'

Freeman Dyson, 1958

pellant. In an Orion ship, some other, inert material can be used – ice, plastic, metal, shipboard waste – anything. In the atmosphere, the air will act as a propellant.

The massive plate would absorb the impact of the explosion, and huge springs would protect the occupants of the ship from being thrown about too violently. 'It will work and it will open the skies to us', Freeman Dyson said in 1958. 'The problem of course is to convince oneself one can sit on top of a bomb and not be fried.' The sight would have been incredible. Normal rocket launches are impressive enough, shattering the air with light and sound. But 300 small nuclear explosions, each half a second apart, would have been truly awesome, turning night into day for hundreds of kilometers around.

Anyone close by would have been blinded and deafened. But the calculations showed that for the crew inside the ship, the launch would not have been too traumatic. According to Freeman Dyson, who still works at the Institute for Advanced Studies in Princeton, the sensation of riding Orion, protected by several hundred tonnes of shielding and powerful springs, 'would have been not that dissimilar to being blasted off in a conventional rocket'. But once in space, on the long silent cruise to Mars, life would be very different to that in today's spacecraft. 'If the Shuttle is something like a motorhome, living on Orion would be more like being in a hotel', he says. 'There would have been a lot of space for exercising, moving around.' No cramped seats as in Apollo or Gemini. No tiny recessed sleeping areas, foldaway tables, and nightmare toilet facilities that are part of life for astronauts now, even on the big International Space Station. In the real world, space travel is all about mass – as little of it as possible. Astronauts are told to keep personal possessions down to a minimum. Spacecraft are constructed from a set of cleverly designed alloys that can both withstand the rigors of orbit and weigh as little as possible. It costs several hundred dollars to send a pencil into space.

But in the Orion world, mass would have been no problem. Quite the opposite. The 4,000-tonne ships would have been constructed of ordinary steel, not lightweight titanium and expensive composites. The astronauts could have brought whole crate-loads of luxuries with them, if they had wished. The ships would have been equipped with comfortable living quarters, showers and toilets, proper kitchens, even a gym. This is because if you are sitting atop a column of exploding nuclear bombs, a lot of weight is actually a good thing, acting as a shock absorber and helping to prevent the whole structure from being blown to bits.

The idea was to send fleets of massive Orion spacecraft initially to Mars, although Dyson himself was particularly keen to go to Saturn. 'We shall know what we go to Mars for only after we get there', Dyson wrote in May 1958. 'You might as well ask Columbus why he wasted his time discovering America when he could have been improving the methods of Spanish sheep-farming. It is lucky that the US Government, like Queen Isabella, is willing to pay for the ships.' Later, he said, 'When I thought about space travel in those days I was thinking about the huge guns that I read about in the stories of Jules Verne. Rockets had nothing to do with it. The Martians in Wells's *War of the Worlds* did not come in rockets. They came in artillery shells.'

In the original Orion Mars concept, two 4,000-tonne cruisers would remain in Mars orbit after flying to the Red Planet at 40 km (25 miles) a second (more than three times the maximum speed of the Apollo craft), while the payload compartment of a third ship, separated from its command module, would land and serve as a surface base. Remember, 45 years after this mission was proposed, the biggest spacecraft we have managed to get to Mars are the size of a family car and do not carry astronauts. And this was just the preliminary Orion mission, a sort of shakedown cruise to test the technology before it was aimed at more ambitious targets in the outer solar system. Freeman Dyson saw no limits to the potential of Orion. Project Deluge would bring huge quantities of water ice, mined from the moons of Saturn, to Mars. With spaceships the size of ocean liners there was really nothing that couldn't be achieved.

The scientists built a test model of Orion – an unmanned craft a meter or so high – and sent it soaring a few tens of meters into the sky, powered not by atomic power but by conventional high explosive. A film exists of the tests and it is quite extraordinary to watch – and quite unbe-

lievable that this could ever be scaled up a thousand-fold and that people might actually volunteer to go aboard. Small the test might have been, but the principle was the same. When Wernher von Braun, initially a skeptic, saw the film of the first Orion test he literally fell off his chair in amazement.

In the end, the US Government pulled the plug on Orion, withdrawing further funding in 1965, after the 1963 Nuclear Test Ban Treaty made the idea of exploding hundreds of atom bombs to get a spaceship into orbit infeasible. Orion was not helped either by the sheer gung-ho craziness of the scientists in charge of the project, who let their enthusiasm for all things nuclear run away with them. As well as exploring the solar system, a civilian concept that had everyone's backing in theory, it was clear early on that Orion had potential military applications – something which, even at the height of the Cold War, was not met with universal enthusiasm in the space and military communities. Huge spacecraft lurking around the Moon would be magnificent platforms for every kind of weapons delivery system imaginable. A scale model, the size of a car, was constructed of a battleship Orion, bristling with cannon and equipped with enough nuclear weapons to raze the Soviet Union from space. In 1962 this true Death Star was shown to President Kennedy who, instead of being enthusiastic, was appalled. 'We have no use for this thing', he muttered, and walked away.

So, Orion died. According to Freeman Dyson, the major technical problem was that he was unable to develop a nuclear device that was 'clean' enough to use on Orion. If conventional atomic bombs had been used, the whole of the launch site and the surrounding hundreds of square kilometers would have been showered with radioactive fallout. Ironically, he is now very happy that he never managed to develop a 'clean nuke'. 'We are very lucky that they did not turn out to be possible', he says, explaining that if they had, then there would have been far less restraint in the past decades when it came to using nuclear weapons.

It seems incredible now that anyone would have thought of using atom bombs for such a purpose. But in the 1950s,

scientists were far more cavalier about using atomic weaponry than they are today. Despite the fact that Hiroshima and Nagasaki were still fresh in people's minds, there was much talk of using atomic weapons for peaceful purposes, such as digging tunnels through mountain ranges and diverting rivers. It is often forgotten that in the 1950s and early 1960s, hundreds of test nuclear weapons, some as large as 50 megatonnes (21,000 times the power of the Hiroshima device), were exploded, above the ground, by the Soviets and the Americans. This was a time before Chernobyl and Three Mile Island, and before the birth of the Green movement.

But the main factor in the demise of Orion was again, according to some scientists, the Apollo program. In 1961, NASA's entire *raison d'être* became the Moon project. Orion would not have been ready in time to meet Kennedy's target, so it was quietly shelved. To supporters of Orion and other nuclear projects, this was a disaster. 'Every available dollar was redirected to Apollo,' says pro-nuclear space commentator Wayne Smith, 'and as a result we now have a flag on the Moon. Nobody has visited our satellite in more than thirty years, and there are no serious plans in the pipeline to do so'.

RUSHING TO THE MOON

Early on in the Apollo planning stage, it was decided that in order to get to the Moon by Kennedy's deadline – 31 December 1969 – it was necessary to get there as quickly and as cheaply as possible. The method chosen was called Lunar Orbit Rendezvous (LOR). In LOR, a Saturn V rocket would blast a lunar spacecraft into orbit. This would consist of a very small landing craft, the Lunar Module (LM), and the larger Command and Service Module (CSM). These fly, attached to each other, from Earth orbit to the Moon, whereupon the LM detaches itself and lands. It then blasts off from the Moon's surface and rendezvous with the CSM, transferring the moonwalkers back on board before returning to Earth. The huge advantage of this plan, devised by NASA Langley Research Center engineer John Houbolt, was that it minimized the amount of fuel that would need to be taken to the lunar surface.

The Houbolt-designed mission got humankind to the Moon. But in many ways it was to have a negative impact on NASA's

(*Opposite*) Since the 1960s, hundreds of design studies have been commissioned for crewed Mars missions. All have foundered on the rocks of economic reality.

post-Apollo ambitions. Because it was so efficient, NASA had no need to develop von Braun's proposed super-Saturn, the Nova, a giant machine that could launch into orbit the amounts of *materiel* needed to get to Mars. An alternative plan – to assemble a Moonship in Earth orbit, was also vetoed on grounds of cost and complexity. This meant that the opportunity to learn how to construct large vehicles in orbit was lost. LOR was a beautiful solution to a taxing problem – how to make the technological leap from what were essentially World War II missiles to a lunar spacecraft in just a decade. But as a basis for interplanetary exploration, it was a dead end. Nevertheless, NASA's engineers kept plugging away with their Mars plans right through the 1960s.

THE EMPIRE PROJECT

In the mid-1960s it was still assumed that some sort of manned Mars mission would take place as a logical follow-on to the Apollo landings. Getting to the Moon, then stopping there, simply was not an option – what would have been the point? Wernher von Braun was, by now, in charge of NASA's Marshall Space Flight Center, and he was keen to build ever bigger and better rockets. He knew that by the time Apollo flew, the work at Marshall in developing the Saturn V and earlier 1B rockets would essentially have been completed. This would have meant the shrinking of the von Braun empire, and to stop this happening he needed to keep the Martian flame alive.

The problem was that, as everybody knew, landing on Mars was going to be extremely difficult. In the early 1960s, no space probes had been there, and little was known of the makeup of the Martian atmosphere, climatic conditions, or hazards that astronauts might face. In 1956, the Italian astronomer Gaetano Crosso demonstrated that a manned spacecraft could visit Mars, and return to Earth, using its rocket only once – to leave Earth orbit. It would coast for the remainder of the flight, and use Mars's gravity to slingshot its way back to Earth. The amount of fuel needed would be more than halved for a landing mission, and total journey time could be reduced to about a year.

The disadvantage of this plan, of course, was that nobody actually got to land on Mars. Instead they would rely on

telescopes to peer at the planet's surface from a considerable distance. The Early Manned Planetary Interplanetary Expeditions (Empire) concept was attacked for its lack of scope. It was pointed out as early as 1963 that if you weren't going to land, and just take pictures from space, you might as well send a robot space probe to do the same job. It was only on the surface of Mars that humans would come into their own. Nevertheless, planning for Empire went ahead, with contracts being awarded for the design phase. It was envisaged that Empire would be carried aloft using a Nova lifter. Lockheed, General Dynamics, and Aeroneutronic all submitted plans.

Of the three contractors, the most ambitious report was compiled by General Dynamics. Its author was a former German tank commander named Krafft Ehricke, who had joined von Braun's team at Peenemunde after the battle of Moscow. His proposed mission, which would take place in around 1975, would consist of a small flotilla of Mars ships travelling in convoy, with the piloted modules having the benefit of artificial centrifugal gravity. This mission would maximize the amount of time spent around Mars: instead of merely swinging by the Red Planet, allowing a couple of days at most for study, the General Dynamics spacecraft would go into orbit, allowing a detailed reconnaissance of the surface to be carried out. The ships would be equipped with probes which would land on the surface of Mars and return to orbit, complete with soil and rock samples. Although the design brief was for a purely spacebound mission, Ehricke's team also factored in a small crewed Mars lander – as the hard work had already been done in getting to Mars, it would be a shame to miss this opportunity. It was only when Mariner 4 made a successful flyby of Mars in 1965 that the idea of a manned mission which did not actually land was abandoned.

Nothing became of Empire, of course. Part of the blame can be laid at the door of Mariner 4, the source of the 'Great Disappointment' in 1965. In just a few hours, Mars was transformed from a possible abode of life with a tolerably thick atmosphere to a cratered wasteland existing in a near-vacuum

(*Opposite*) By the late 1960s, it was increasingly apparent that any future manned exploration of Mars would be undertaken in conjunction with a continued robotic program.

'You've just seen something very extraordinary. You've just seen the death of the planetary program'

Jack James

where the possibility of life was effectively nil. As David Portree writes, 'The Mariner 4 results eradicated any lingering traces of Lowellian romance and in fact shifted the prevailing view of life on Mars all the way down the continuum to a pessimism with almost as little basis as Lowell's optimism.'

An anecdote from the Mariner 4 mission gives an insight into the gathering storm that was about to flatten America's hopes for space exploration. As the little space probe swooped past Mars, it sent its data back to Earth, bit by bit, via a high-gain transmitter. Unfortunately, the software needed to convert this data into a useable image – a picture of the surface of Mars – kept on crashing. This is almost unimaginable now, in the days when a $300 PC would be able to do the job in a tenth of a second, but back then the total computing power available to NASA was probably less than is contained in the machine on which this book is being written. Jack James, the lead engineer, and his assistant decided that, computer or no, they would have an image, and began to color in a sheet of squared graph paper by hand with a pencil – the shading of each square, or pixel, being determined by the hexadecimal value of each byte of data. It was a laborious process, to say the least. One can imagine the look on the two scientists' faces as the picture slowly started to take shape. Instead of the canals, forests and cities that some were still half-expecting to see, the graph paper started to fill with circular craters. What they ended up with was a picture of what looked very much like the Moon. According to colleagues who witnessed this event, James turned to his assistant and said, 'You've just seen something very extraordinary. You've just seen the death of the planetary program.' And then he added, 'We've got about an 8-hour head start on everybody else in looking for a new job.'

After the murder of President Kennedy, Lyndon Johnson proved to be as enthusiastic as his predecessor about Apollo, but his keenness for a Mars mission was limited. Vietnam was beginning to soak up a lot of money, and the billions of dollars a crewed Mars landing would cost would be hard to justify. By the late 1960s – before even Apollo 11's triumph – NASA budgets were being cut for the first time. In January 1967 a fire on board an Apollo spacecraft on the ground killed three astronauts in the most horrible way, dealing a severe blow to NASA's hopes and ambitions. The politicians started to demand more and more 'oversight' of the Agency's aims and ambitions – and in particular its spending plans. Mars was starting to slip over the horizon of what was considered financially possible and publicly acceptable.

Chapter 6
Not Going to Mars

'Basically, we don't have the money'
Congressman Robert Traxler (1990)

By the late 1960s, the space race was in the home straight, and America was so far in the lead that as far as the Soviet Union was concerned, it was game over. The suggestion that by 2004 humanity's presence in the cosmos would have been reduced to an expensive and – many argue – pointless space station in low-Earth orbit would have been risible.

By now, even taking into account the worst-case scenarios, the Vietnam War and its appetite for cash, and the trials and setbacks that come with any exploration project, humankind

If they ever get there, astronauts will be forced to spend a long time on the surface of Mars. They will have the opportunity to explore far and wide, and most mission design studies have incorporated some sort of motorized vehicle.

should at least have established a permanent presence on the Moon, and have set foot on the planet Mars. In 1969 this was taken for granted by everyone, from politicians and bean-counters to rocketeers and scientists. From the perspective of those brave days, the future is one in which space exploration has, puzzlingly, stopped. But from the perspective of the early twenty-first century, the optimism of the Apollo era, while infectious, can be seen to have been built on rather a lot of shifting sand. Going to the Moon had been predicted by science fiction and we did it – just. Going to Mars is also a staple of science fiction, but Mars is, even at its closest, 150 times farther away than the Moon and probably a thousand times harder to get to.

Maybe the current hiatus in the space program is a good thing, as it will allow us to redefine our priorities and develop technologies that will make space exploration more reliable and safer. This is one justification for the International Space Station – if nothing else, it has allowed us to develop space technologies that demonstrate long-term reliability. But looking back on those heady days when humans were set to conquer the solar system, it is hard not to be struck by a tremendous feeling of nostalgia – nostalgia both for a past where anything seemed possible, and for a future that never was.

NASA continued with low-key plans to go to Mars right through the 1970s and after. The Voyager plan included sample-and-return probe to bring possibly biological material from the Martian surface to the hands of trained biologists in Earth laboratories. Eventually, an ambitious series of Voyager missions was to lead to a manned mission, probably a flyby of Mars. But Voyager (not to be confused with the later, highly successful mission of the same name to the outer solar system) died in August 1967, the victim of spiraling budget deficits and the Vietnam War. NASA's budget was trimmed by more than half a billion dollars, and manned interplanetary missions were right off the agenda. Voyager mutated into the Viking missions of 1976. Getting humans to Mars seemed as far away as ever.

However, the dream wouldn't die. Throughout the late 1960s, and into the 1970s and beyond, various individuals, companies, and organizations continued to make plans to get humans to Mars. In January 1968, Boeing proposed what was probably the largest and most ambitious Mars ship since the days of von Braun's fanciful behemoths. Boeing's idea was for a gigantic 30 m (110 ft) craft to be constructed in Earth orbit and powered by an even more colossal 150 m (500 ft) propulsion assembly composed of five separate modules. The whole construction would weigh around 2,000 tonnes. Motive power would come from a NERVA-like nuclear engine which would eject the 1,000 tonnes or so of hydrogen propellant contained in the propulsion module tanks. Boeing calculated that the Mars ship could be launched into space aboard just six uprated Saturn Vs, each some 150 m tall (the originals were nearly 50 m/150 ft shorter) and boosted by strap-on solid fuel boosters.

Depending on the scale of the mission (which meant the scale of the funding), the astronauts – either a two-man or four-man team – would remain on the surface of Mars for anywhere between four days and a month. Boeing put the total cost of a manned Mars mission, at late-1960s prices, at about $30 billion – close to the quarter of a trillion dollars a Mars mission is still estimated to cost today.

It had no chance. The Johnson administration had little enthusiasm for such grand plans. Interest in space was on the wane, and talk was of budget cuts at NASA, not expansion. The pessimism was noted by *Aerospace Technology* magazine, which predicted in May 1968 that the proposal was likely to be 'the last of its type for a year'. In fact, it was a lot bleaker than that: there would be no more serious, NASA-funded mainstream proposals from major contractors to put a person on Mars until the late 1980s. NASA's budget was now in freefall. The NERVA project looked doomed, as did plans to build more Saturn V rockets.

In 1969, Thomas Paine became the new administrator of NASA, and Richard Nixon became the new US president.

Paine, a Democrat and a supporter of a Mars mission, fought a long battle with the new administration. Although the Boeing monster was a non-starter, other plans were put in place for NASA's future which included manned Mars missions. The Space Task Group was a political exercise by Nixon to examine possible futures for NASA; in reality, it seemed to turn into a cost-cutting committee. Nevertheless, the politicians paid lip-service to the idea of Mars. At the Apollo 11 launch, Vice President Spiro Agnew, a fan of space exploration, talked of the 'simple, ambitious optimistic goal of a manned flight to Mars by the end of the century'.

The problem was that the public was not interested. A Gallup Poll taken in 1969, after the successful landing on the Moon by Armstrong and Aldrin, showed that it was only among the under-thirties that there was any enthusiasm for an American manned Mars mission. A large majority of older people opposed such a mission. Overall, less than four out of ten Americans favored even the idea of a Mars mission. Interestingly, the polls give the same picture today.

The scientists hit back. They stopped talking about space in terms of exploration and dreams, and instead started listing the potential benefits that the new technologies could bring. The space program was now talked up in terms of new alloys and amazing materials which would soon be in every kitchen. It became received wisdom, for example, that the nonstick coating with the tradename Teflon was a direct result of the Apollo program; in fact, it wasn't. Apollo did, though, generate many new technologies. The ferocious need for computing power led to the development of miniaturized transistors and light-weight computers that in the end gave us the microchip. The appetite for power to drive the myriad systems and subsystems aboard the spacecraft led to the development of fuel cells; if we really are driving around in clean, green hydrogen-powered cars in twenty years' time, we can thank Apollo. Even something as humble as the cordless power tool was a child of the lunar landings: the astronauts were equipped with rechargeable rock drills to take core samples from the lunar basalt.

But the tactic of justifying space by listing the economic benefits can be seen, in retrospect, to have misfired badly, and may be one of the major reasons why the manned space program stalled and remains so to this day. While new alloys and even medicines are interesting, they hardly fire the enthusiasm of the public in the way that spectacular missions to the Moon and the planets do. Making space all about technology spin-offs and industrial benefits sounds very worthy, but it is, like many very worthy things, in essence dull. Non-stick frying pans and electric drills just aren't as exciting as a flag on Mars.

And bad news was on the way – not from Congress or from the bean-counters, but from Mars itself. Almost unnoticed amid the Apollo hullabaloo, Mariners 6 and 7 arrived at Mars and started sending back reams of data – high-quality photographs, showing more craters and mountains, and measurements indicating a thin atmosphere. For many people, this pulled the rug out from under those agitating for a manned Mars mission; the rug had already been given a sharp tug by Mariner 4.

Nevertheless, Wernher von Braun, the great rocket salesman, was convinced that he could persuade the politicians that humans should visit Mars. On 5 August 1969, he gave a talk to the Senate Space Committee outlining the huge scientific payoffs from a Mars mission, which he thought he could mount by 1981. His ambitious Mars plan consisted of two ships – each constructed from three nuclear spacecraft and a mission module; the whole vehicle was 30 m (100 ft) across and more than 60 m (200 ft) long, to be constructed in orbit after the constituent parts were carried into space aboard a series of Saturn Vs.

Von Braun set out very clear objectives for the astronauts on their arrival at Mars. This was to be no flag-planting exercise. A high priority was to be the search for life – von Braun believed, as did many other people, that life, even higher forms of life, may exist on Mars. He also thought it vital that the visitors search for water, which would 'open many possibilities'. He thought it might be possible to use any Martian water to produce rocket fuel for future missions. How much would all this cost? Von Braun told the politicians that NASA would swallow up a mere 0.6 percent of America's Gross National Product (about $7 billion by the mid-1970s) with the amount as a proportion of the nation's wealth declining after that.

The Senate rejected the plans, citing cost as the main objection. Despite this, NASA went ahead and prepared a report outlining future plans to the newly formed NASA watchdog, the Space Task Group (STG), which was under the chairmanship of Vice President Spiro Agnew. 'America's Next Decades in Space: A report to the Space Task Group' was presented on 15 September 1969. It was a long plea to maintain the momentum of Apollo, and not acquiesce to those who were beginning to say, when it came to space exploration, 'Been there, done that.' The report outlined von Braun's plans, and gave three timescales for an assault on Mars: a money-no-object 1981 launch, a delayed 1986 launch and a non-specified launch date. In turn, the STG produced its own report, largely based on the NASA document but with rather different conclusions. Rather than a firm commitment to Mars exploration, there were vague promises of a manned landing by the end of the century. Agnew, for his part, continued to fight a losing battle for Mars, even though he was actually booed in public several times when he spoke of it. When he insisted on including a mention of a possible 1981 landing in the STG recommendations, he was hauled in to see Nixon. After 15 minutes in the Oval Office, he emerged to announce that the Mars mission was no longer 'recommended' but merely 'technically feasible'.

Nixon killed off manned planetary space exploration in favor of a far more limited program to develop a reusable spaceplane (the Shuttle) and perhaps an orbital space station. NASA's budget was cut to $3.5 billion, and on 13 January 1970 administrator Thomas Paine announced the closure of the Saturn V production line and the cancellation of Apollo 20 – the tenth lunar mission – in favor of Skylab, which would use the last Saturn V.

Paine resigned, and NASA was subjected to further attacks. It was decided that the way forward was the Space Shuttle, which, it was reasoned, would be a far cheaper and more effective way than using the Saturn V of getting equipment into orbit. A final study was commissioned for a manned Mars landing, a chemical rocket-powered version of Boeing's original behemoth, but it came to nothing. Then NERVA was cancelled in 1973. By the end of 1973, manned space flight meant low Earth orbit – where it has remained ever since. In fact, NASA was lucky to maintain even this capability. In July 1971, Congress moved twice to delete totally the Shuttle from the 1972 budget. Only exaggerated claims of how many jobs would be generated by building the Shuttle, announced in critical states in the 1972 election year, saved the spaceplane.

Of course, many at NASA were horrified at the emasculation of the manned spaceflight program. Harrison Schmitt, the penultimate Moonwalker and the first and only scientist

ever to have set foot on the Moon, triggered a renewed flurry of interest in Mars during the late 1970s. In 1985, President Reagan launched the National Commission on Space (NCOS), the successor to the STG. It was headed by Thomas Paine, the former NASA administrator who had resigned in the wake of Nixon's budget cuts. The NCOS report was full of talk of lunar bases and setting up a Mars colony by 2026. It was shelved. A further report by Shuttle astronaut Sally Ride, which also included plans for a landing on Mars by 2005 (she was probably the first senior NASA figure to state that no human would walk on the Red Planet before the twentieth century was up), attempted to maintain interest in solar system exploration while taking into account budgetary realties.

THE SPACE EXPLORATION INITIATIVE

All presidents want something to be remembered by. Kennedy's historic announcement back in 1961 that the United States would land a man on the Moon by 1970 was just the sort of thing to guarantee any White House resident a place in the history books, whatever the slings and arrows their administration was forced to contend with. This was what the newly elected George Bush Sr thought in 1989, shortly after he was elected. The result was the Space Exploration Initiative, perhaps the most grandiose and impressive space exploration plan to see print since the Apollo era. If the SEI had been implemented in full, we would now have a very large and quite possibly useful space station in low Earth orbit, we would have a semi-permanent presence on the Moon, and we would be deep into the serious planning for a manned Mars mission. The SEI failed, spectacularly. It was, in fact, laughed out of Congress. But it deserves to be remembered as a final attempt to recapture the glory days.

In his speech on the steps of the Air and Space Museum in Washington to mark the twentieth anniversary of the Apollo 11 landing, Bush promised that NASA would go 'back to the

(*Opposite*) As the years have passed, putative Mars ships have got smaller, sleeker and more refined, a world away from the behemoths of Wernher von Braun. They are still no closer to being built, however . . .

'back to the Moon, back to the future, and this time back to stay . . . and then, a journey into tomorrow, a journey to another planet, a manned mission to Mars'

President Bush (sr)

Moon, back to the future, and this time back to stay . . . and then, a journey into tomorrow, a journey to another planet, a manned mission to Mars'. Those words were immediately splashed across newspapers and TV screens. But it soon became clear that SEI wasn't going anywhere, and it became something of a joke.

The rough timetable was that by 1995 the new space station, then to be called 'Freedom', would become operational. In the first decade of the twenty-first century, after a series of robotic explorations, a manned lunar outpost would be set up. Planning for a Mars mission would begin straight away, culminating in a mission to be launched in the second decade – perhaps in 2017 or 2019. The hope was certainly that by the fiftieth anniversary of the first lunar landing, the surviving moonwalkers (if there were any) would be able to commentate in the TV studios on the spectacular pictures of an American boot treading the dusty red soil of Mars. It was a small speech, but the intention was to deliver a giant kick up the backside of NASA and the entire space community, give America some purpose and vision, and set Earthlings once more on course for the stars.

NASA was given the job of turning these vague outlines into a concrete plan for its future. Deputy Administrator Aaron Cohen led a team of some 160 NASA managers, engineers, and scientists in drawing up one of the seminal documents in the Agency's history, the 90-Day Study. The Study concluded that NASA's future lay in constructing the space station Freedom, and returning to the Moon and thence to Mars. The Study outlined five different timelines – with a Mars landing taking place anytime from 2011 to 2016.

On close reading, the Study turned out to be a shambles. According to the respected space policy commentator Dwayne Day, who published one of the most recent retrospective accounts of the SEI,

Without strong leadership from Headquarters to establish clear priorities, the 90-Day Study essentially became a shopping list of every program that various NASA constituencies wanted, regardless of whether they were necessary to achieve the new space policy and regardless of how much they would cost.

On this shopping list were items like lunar robotic probes, and a massive network of Mars exploratory probes including a $10 billion sample-and-return mission. There was to be a permanently occupied lunar base and a Mars base, and the development of a new heavy-lift booster based on Shuttle architecture, called Shuttle-C.

Some of the items were frankly bizarre. One of the requirements for the lunar base was a crane for lifting cargo off the landers and depositing it on carts to be wheeled to the base. The price of the crane was $10 billion. There was also a requirement for fourteen Shuttle flights a year – no one explained how this was possible when the fleet could manage no more than nine. The cost analysis was suppressed almost instantly when shocked officials saw the figures. In the end, it was understood that the bottom line for a return to the Moon and a manned landing on Mars was around half a trillion (1991) dollars.

That's a scary sum, but it needs to be put into perspective. The US defense budget for 2003 was set at a shade under $400 billion. Over the quarter-century timescale that the SEI addressed (the cost analysis was a *total* for the period 1991–2025), America would spend around $10 trillion or more on its military – one hundred times the total cost of a mission to Mars. The price of going to Mars, to the average American taxpayer, would have worked out at around $60 a year. This compares quite favorably with the cost of Apollo, which at full steam was costing each and every American the price of a packet of peanuts a day.

But with a $3 trillion budget deficit, a space program that was rumored all over Capitol Hill to cost a quarter to half a trillion dollars – even spread over more than a quarter-century – was a political disaster. The United States was in no mood to increase its overdraft. Bush quietly began to backtrack, killing off SEI not in a single flourish, but strangling the beast slowly and painfully. He requested $216 million to get SEI started, but even that was knocked back by Congress.

In 1993, the Office of Exploration's request for an extra $5 million – paperclip money in NASA – was struck down by Congress, and bean-counters in Washington combed through NASA's budget and deleted programs even if they only sounded as if they were part of SEI.

Representatives from both parties shot SEI down in flames. Robert Traxler (Democrat, Michigan) put it most bluntly: 'Basically, we don't have the money.' The president did not allow himself to believe that his vision was entirely dead. As late as 11 May 1990, he made a speech in which he restated his hope that an American flag would fly on Mars in time for the fiftieth anniversary of the first Moon landing. On 20 June he said he would continue to fight for Mars. Less than ten days later, Congress killed off SEI outright, voting to give NASA not a single cent towards projects to return to the Moon or to go to Mars.

NUCLEAR POWER? YES PLEASE

The humiliation of George Bush Sr is not something that any president would want to replicate – least of all the current incumbent, his son. Perhaps this is why George W. disappointed space enthusiasts everywhere when he made a speech at Kitty Hawk, North Carolina on 17 December 2003. The US media had been predicting for some weeks that the president would use the occasion, a celebration of the centenary of the first powered flight by the Wright Brothers, to make a grand announcement about space travel – something along the lines of his father's 1989 address on the steps of the Air and Space Museum, perhaps 'We will go to the Moon, this time to stay.' But in the end there was nothing – no Moon, no Mars. Nevertheless, less than a month later Bush indeed came up with the goods. (His speech, and its implications for future travel to the Moon and Mars are discussed in Chapter 14.)

The key to any revival of plans to go to Mars may lie with atomic power. In fact, the idea of atomic rockets has never entirely gone away. Perhaps sensing that the Green movement has lost some of the clout it acquired in the 1990s, nuclear supporters are once again stalking the corridors of NASA, and now the Agency is revisiting the spacenuke. Realizing that exploring space properly is going to need this type of technology, America's space chiefs are reopening research into atomic-powered spacecraft.

MARS DIRECT: GETTING THERE ON THE CHEAP

There have been almost as many plans for getting to Mars as there are craters on its pockmarked surface: nuclear ships, giant Saturns, Orions, fleets of interplanetary cruisers. The technologies vary, and the dreams behind these missions are different. But one thing has remained constant in mission planning for the past forty years: going to the Red Planet will be a huge undertaking, swallowing up a significant percentage of the GDP of even the United States – costing, in fact, as much as a small war. But one mission plan has stood out from all the others for its apparent simplicity, its avoidance of the need for gargantuan, heavy-lift rockets, its reliance on tried and trusted technology, and, most pertinently, its cost. One man, Bob Zubrin, believes that not only can we get to Mars more or less right away, but that we can do it for just $20 billion – chickenfeed when it comes to manned space exploration. His plan is called Mars Direct, and elements of his plan will almost certainly be incorporated into the first Mars missions.

Mars Direct originated in the early 1990s, and was strongly influenced by the study headed up by astronaut Sally Ride. It incorporated some key technologies missing from other Mars mission plans, most notably the capability to manufacture rocket fuel on the Martian surface. This alone would slash the cost of the mission, Zubrin believes. Because it is possible to break down carbon dioxide into methane and water, then, provided liquid hydrogen is available, you can in theory mine rocket fuel from the Martian atmosphere. Just 5 tonnes of liquid hydrogen feedstock would be enough to manufacture more than 100 tonnes of rocket fuel in one year, by a complex series of reactions, turning CO_2 into methane and water, and then turning the water into hydrogen and oxygen, recycling the hydrogen to produce more water and methane. More oxygen could be made by turning CO_2 into CO. All this would be carried out by a large, robotic lander-factory sent to Mars aboard a Shuttle-derived heavy-lift rocket which could get to Mars, direct, in six months. What would arrive on Mars would be 40 tonnes of *materiel*, including the fuel/oxygen factory, a 100 kW nuclear reactor, and an Earth Return Vehicle to carry the astronauts back home. There would be no need to assemble a large Mars cruiser in orbit, hence the 'direct' part of the mission.

Three years later, two more ships would be launched to Mars. One would be identical to the first robotic fuel-plant. The second would include a large manned lander, capable of supporting a crew of four, packed with supplies of food, water, and propellant, and with enough storage space for an electrically powered, pressurized rover. The lander would stay attached to its upper-stage rocket by a steel tether, and the whole assemblage set to rotate to produce artificial gravity equivalent to that on Mars. In this way, the astronauts would avoid the physiological trials of weightlessness. Both ships would arrive more or less simultaneously, the piloted ship discarding its centrifugal counterweight and plummeting down to where the original robot lander arrived three and a half years previously. Then the crew would have to rendezvous with their supply of fuel and with the rocket to get them home. Ideally, they would land within walking distance, although if they were off-target they could use the rover to get there. The worst-case scenario would be a landing more than 1,000 km (600 miles) from the fuel plant. If that happened, the crew would have the option of commanding the cargo lander, launched with them, to land nearby. If this lander failed there would be sufficient supplies aboard the expedition ship for the explorers to hold out for the two years necessary for a relief crew to arrive. But if all went well, the later cargo lander would put down some 800 km (500 miles) from the crew, establishing another supply-base for the next Martian expedition. A series of repeat launch cycles would, over time, establish a network of what Zubrin calls 'towns' on the Martian surface. Mars Direct would, Zubrin has been arguing for a decade now, provide NASA (or anyone else) with a cheap, reliable way of getting humans to Mars.

But would it work? The concept of ISRU (In-Situ Resource Utilization) is very attractive. Humans are superb exploring machines, but unfortunately they each requiring more than a tonne of fuel and lubricant to survive an extended Martian voyage. Humans cannot directly utilize solar power nor radioactively generated electricity. Making water and air on Mars would solve many if not most of the problems.

But the trouble is that no one has yet built the sort of machines Zubrin is proposing. In theory we know how to make methane and water from Martian raw materials, but no one has done anything like it in practise. According to Jim Garvin, NASA's chief scientist for Mars exploration, Mars

Direct looks far better on paper than it does in reality. 'Zubrin's ideas are innovative but naive,' he says:

> We have no technology to realistically accomplish the ISRU that Zubrin requires, and his views of the Mars we are exploring with Spirit and Opportunity are old and already out of date. We do not have the flight systems he is hawking, and development times have never been as short as he touts.

THE REFERENCE MISSION

From the time of the SEI debacle until now, the main paradigm for Mars mission design has been the Mars Direct concept, originated by Martin Marietta and championed by Bob Zubrin. In 1992, NASA held a Mars Exploration Study Team Workshop which produced a mission concept that became known as the Design Reference Mission. The Reference Mission contained seven key elements. There would be a large, heavy-lift rocket capable of launching 240 tonnes into low-Earth orbit, launching 100 tonnes into Mars orbit, and delivering 60 tonnes to the Martian surface. There would be a crew of six, no reliance on Earth orbit assembly or the Space Station to construct the Mars ship, and reliance on ISRU technologies. The main difference between the original Mars Direct plan and the DRM was the use of Mars orbit rendezvous – a smaller landing craft would lift off from the surface and dock with the Earth return vehicle.

The crew would be equipped with a pressurized rover vehicle. As with Mars Direct, there would be a 'base camp' set up on Mars to manufacture fuel before the astronauts arrived, and the mission design would allow a series of expeditions to be mounted, each using hardware carried to Mars by previous launches. Thus a cyclical series of explorations

The Reference Mission of the 1990s was a design for a slimmed-down assault on Mars using a small crew and several of the 'live off the land' ideas of Bob Zubrin and others. Mars has the raw materials necessary to make rocket fuel, and several experimental instruments have been built which demonstrate the feasibility of extracting and synthesizing chemicals such as methane and hydrogen from the Martian atmosphere. But some doubt whether such technology is sufficiently reliable to form the basis of a crewed mission in the near future. When, and if, a Mars mission is eventually mounted, the returning astronauts will be feted the world over.

(Opposite) Pioneer explorers will have to bring with them everything they need. 'In situ resource utilization', if it could be made to work, would greatly increase the viability of Mars as a place to live. If you can make air to breathe, water to drink, and rocket fuel to get you home, then going to Mars becomes a lot easier. But on the surface you will still depend totally on pressurized, radiation-shielded habitats.

of Mars could be conducted, taking advantage of the roughly two-yearly oppositions and launch opportunities from Earth.

Although interest in Mars grew during the 1990s, largely as a result of the discovery of possible fossilized Martian bacteria in the ALH 84001 meteorite, it became clear that the original DRM plan was not going to work – mainly because it relied on the development of a wholly new heavy-lift launcher. The development of this vehicle alone would have sent costs spiraling way beyond NASA's budgetary constraints. In 1997, a slimmed-down version of the Reference Mission was published which eliminated the need for a heavy lifter and instead used some eight launches of a Shuttle-derived launcher capable of carrying 85 tonnes into space. Before the recent sea change at NASA announced by President Bush in 2004, the DRM was the last gasp for those hoping to see a human on Mars in the foreseeable future.

THE INTERNATIONAL SPACE STATION

I had only been in Moscow for about three hours, and as my time in the Russian capital was extremely limited, I decided to take in some sights. Shortly before 12 p.m. on a very cold November night back in 1998, I found myself standing next to Lenin's tomb. I noticed a group of people standing close by, speaking English, and asked them to take my picture. It turned out they were in Moscow on space business, as was I. My mission was to cover the launch of the first module of the International Space Station (ISS) for my employer, a British newspaper. Their mission was ... well, I never found out what their mission was, because they clammed up sharply when they worked out that I was a reporter. I did discover that they worked for NASA. I also found out their opinion of the machine the ISS was meant to replace, the ill-fated Mir Space Station. 'Piece of shit,' one said, 'sooner they dump it in the ocean the better as far as we're concerned.'

Twenty hours later I stood, shivering, on a godforsaken steppe about 2,000 km (1,250 miles) south of Moscow, watching a Proton rocket haul the Zarya module into the leaden, Kazakh skies. An hour later we were all bussed to a brand-new building in the middle of the town of Baikonur. Actually, we weren't in Baikonur at all – Russia's Cape Canaveral was,

so the legend goes, named with deliberate confusion in mind, to confound the Yankees. The real Baikonur stood a couple of hundred kilometers to the north. We were in a settlement that really should be called Leninsk (which even more confusingly was built up around a railway junction called Tyuratam), which is the HQ for the Russian launch program. The vodka flowed, and tongues were loosened. One cosmonaut owned up to smuggling large quantities of vodka onto the Mir station, and predicted ruefully that such transgressions would no longer be possible on the new ISS. But what I remember best was this: another spaceman (who shall remain nameless) announced loudly and clearly, 'We are on our way to Mars.'

The story of the ISS is long and convoluted. In summary, the station was conceived at the height of the Cold War to be an American outpost in space. It was christened Space Station Freedom, then Alpha, and was a favorite project of Ronald Reagan's. When Cold War tensions started to ease, the station was redesigned, seemingly several dozen times, with each reincarnation being less ambitious than the last. Finally, amid much hoo-ha, the ISS was born, a sort of bastardized Mir, bigger and better than the old Russian station but of the same fundamental design. The first planners of the super-station believed that an orbiting platform would be useful as a construction site for interplanetary craft – something envisaged as far back as the 1950s.

The other way in which the new station was supposed to help us on our way to Mars was to act as a testbed for extremely long crewed missions. Getting to Mars and back was going to take at least three years, possibly longer. It was not known whether human beings could survive the two years plus in zero gravity that a Mars mission might require – hence the need for a large, orbiting crewed platform. Back in 1948, Wernher von Braun understood the psychological problems that might arise on such a long mission. He wrote:

(*Opposite*) The International Space Station (ISS) is either 'Mankind's Embassy in Space', as its supporters insist, or a $100 billion white elephant devoid of purpose and responsible for sucking valuable cash away from more worthy and interesting projects, such as the exploration of Mars. It is hard to argue that the ISS, which is currently capable of supporting only a skeleton crew (missions involve just two astronauts), is advancing the cause of space exploration or science. Still, for the moment it is all we've got.

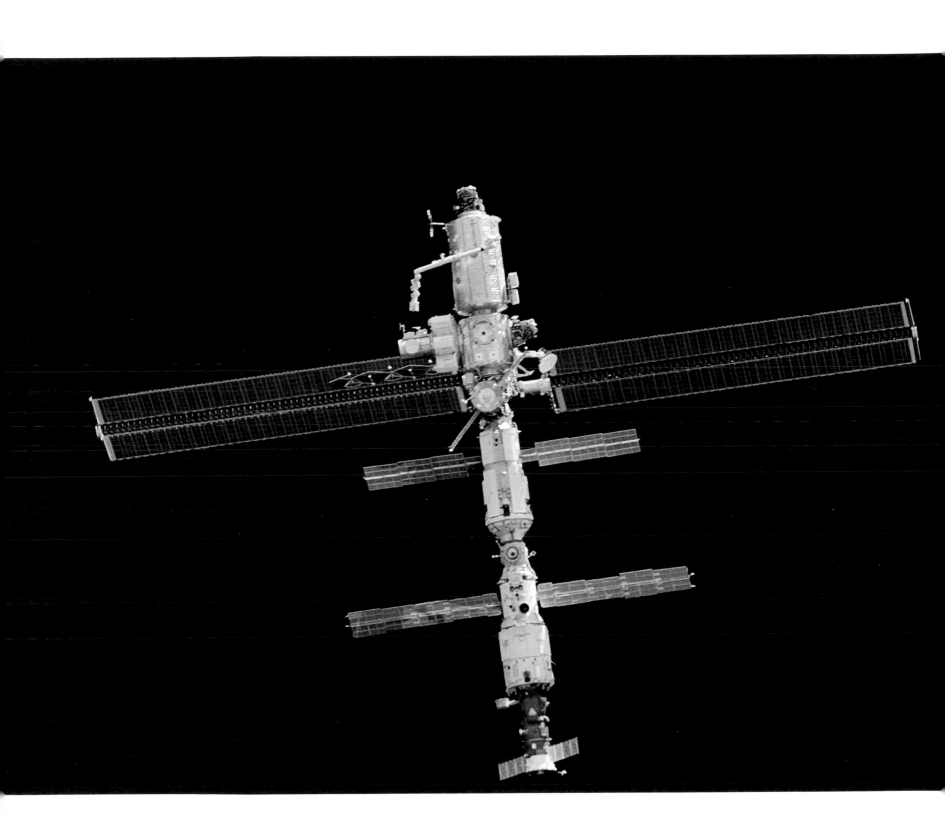

At the end of a few months, someone is likely to go berserk. Little mannerisms, the way a man cracks his knuckles, blows his nose, the way he grins, talks or gestures – create tension and hatred which could lead to murder. If somebody does crack, you can't call off the expedition and return to Earth. You'll have to take him with you.

Most dramatically, von Braun advocated total censorship of radio communication from Earth to prevent the crew learning of any bad news concerning their families or hometowns.

Supporters of the ISS claim that without it we could not learn how the body reacts to long periods in space, yet it can be argued that previous generations of Soviet space stations answered many of these questions. On the Salyuts and on Mir, cosmonauts had spent progressively longer and longer times in space. The record, to this day, is held by Dr Valery Polyakov, who spent 438 days aboard Mir in 1994–95 – long enough (just) for a return trip to Mars. The Russians, who lost the race to the Moon, had been quietly researching the psychology and physiology of a crewed Mars mission for nearly three decades aboard their increasingly sophisticated space stations. Polyakov probably has more insight into what a Mars mission would mean for the crew than anyone else on Earth. When he broke the record he was in his fifties, and to this day he maintains that any crew to Mars should be composed of men and women aged fifty to sixty, or even older. 'This is the age when a person reaches a peak in his or her physiological, psychological and social stability', he told *Space News* in 2000. 'Besides, interplanetary travel might involve an unknown risk for [the] reproductive systems of cosmonauts.' For their part, supporters of the ISS maintain that the station performs a dual function – that of testing both human bodies and spacecraft systems.

WILL WE EVER GO?

In the 1960s, it would have seemed incredible that Apollo would be a one-off. In the eye of the public, space travel would follow a logical progression: first the Moon, then the establishment of lunar bases and perhaps the building of giant space stations. Then, inevitably, would come the conquest of Mars and perhaps the outer solar system as well. The timescale proposed in Stanley Kubrick's *2001: A Space Odyssey* did not seem so ridiculous in 1968; after all, if a landing on the Moon was less than a year away, nuclear rockets were in the pipeline, and Mars would be a destination by the mid-1980s at the latest, the idea of a manned mission to Saturn by the beginning of the twenty-first century seemed quite logical.

In fact, Apollo was always going to be a one-off because of the context of its birth. The American space program would probably have existed without the Soviets, but nothing like NASA would have been created so quickly. Sputnik, then Yuri Gagarin's first spaceflight, are seen now through the rosy lens of nostalgia, and it is easy, in an era when the United States is the only superpower survivor, to forget the shock and horror the Soviet achievements engendered in America at the time. In his highly entertaining biography *Rocket Boys*, (made into the film *October Sky*), the former NASA engineer Homer Hickham recalls how he and his friends, even in darkest West Virginia, were inspired by the early successes of the Soviet space program to study mathematics and science, with the intention of helping America to keep up. In a memo to Vice President Johnson, Kennedy made it clear what his priorities were:

> Do we have a chance of beating the Soviets by putting a laboratory in space, or by a trip around the Moon, or by a rocket to land on the Moon, or by a rocket to go to the Moon and back with a man? Is there any other space program which promises dramatic results which we could win?

In an article for *Spaceflight* magazine in 1996 entitled 'How NASA Lost the Case for Mars in 1969', the British space writer Stephen Baxter commented:

> So when Kennedy made his famous May 1961 commitment to reach the Moon within the decade, the new programme was not seen, by the politicians, as a first step in an orderly expansion into space. Rather, Kennedy was reacting to several factors, among them the early Soviet lead in spaceflight and his administration's Bay of Pigs disaster [the failed, shambolic invasion of Cuba]. Apollo was seen, politically, as an end in itself, with a precisely defined political goal: to beat the Soviets into space.

Science fiction writers such as Arthur C. Clarke like to imagine a future in which international cooperation opens up the possibility of serious space exploration. History shows that the reality is rather different. At the height of the Cold War, some half a million Americans were employed in the space program – that is an incredible one-quarter of 1 percent of the entire population. NASA swallowed some 5 percent of the federal budget. Now the numbers are in the tens of thousands and dropping. It is quite possible that had the Soviet Union never existed, or if it had neither the expertise nor the resources to challenge the US, then humans would not to this day have walked upon the Moon. It is also possible that had the Soviets won the race to the Moon (a not entirely fanciful possibility), this would have spurred the Americans to go one better and head for Mars (the Soviets had many drawing-board plans for Mars missions, but their backs were broken after the failure of the Soviet manned Moon program). Now all is peaceful between the Americans and the Russians, and the result is the International Space Station. The ISS itself is the result of

political calculations – calculations which concluded that it is better for America to bankroll Russia's financially decrepit but technologically capable space program and give it something to do, rather than have all those talented rocket scientists going off to work for someone nasty in a cave in the Middle East – specifically (at the time) Iraq, Syria or Libya. The desire to prevent Russian technology falling into the wrong hands has obviously intensified post-September 11.

It will almost certainly be America that goes to Mars – leading the way, perhaps, as part of a large international team that will have to include the Russians and Europeans and, who knows, maybe even the Chinese as well. In Chapter 14 the latest plans for a manned mission to Mars are discussed – plans that read worryingly, to some, like the Space Exploration Initiative of 1989. Even if the plans outlined in the latest presidential announcement come to fruition, it is unlikely that we will see a crewed mission to Mars much before the third decade of this century, possibly later than that. In the meantime, exploring the Red Planet will have to be left to the robots.

Chapter 7
Invasion of the Robots

'Faster, better, cheaper . . .'
Daniel Goldin

In the 1890s an obscure Russian schoolteacher and amateur scientist named Konstantin Tsiolkovskii worked out that only the sustained thrust of a rocket motor could make space travel a reality. He predicted with fair accuracy the essential elements of all the space missions of the century to come, including multi-stage liquid-fuel rockets that use a succession of engines which are cast away when their fuel is spent. But it was not until the advent of the liquid-fuelled multi-stage rocket by Robert Goddard in the 1920s that the practical way forward for possible space exploration became clear. By 1944, German rocket-propelled missiles were pounding London; by the late 1950s both the Soviets and the Americans were attempting to fire probes at the Moon, eventually reaching their targets successfully. The stage was set for an assault on Mars. But what would these probes find?

For nearly forty years, a succession of probes have tried – and mostly failed, it must be said – to reach Mars. Taking advantage of the 26-monthly launch windows that give Mars-bound probes the most fuel-efficient trajectory from Earth, there will be a probe entering Mars orbit or landing on its surface every two years or so for the foreseeable future. From the probes which have succeeded, we have learned an extraordinary amount. The Martian surface has been mapped with a precision that would have astonished the nineteenth-century cartographers of our home planet. The Viking orbiters of 1976 produced humanity's first large-scale atlas of another planet. The twin landers which were part of this mission took the first photographs from the surface, and brought Mars, as a place, into millions of living rooms. There are now probes orbiting Mars which can take photographs of objects only a couple of meters wide; in a few years, a new generation of

electronic eyes should be able to see things as small as a coffee table or even a dinner plate. By the end of the 2020s we may know the face of Mars better than that of any other object in the solar system. At time of writing, two extraordinary wheeled robots are trundling around Martian locations which appear to have once been home to water. In a few years, other robots will land, scoop up soil and rocks, and bring them back to Earth.

With Mars right in the center of America's sights, the pace of robotic exploration will accelerate. There are now plans to send people to the Red Planet sometime in the 2020s or, more likely, the 2030s. A full, robotic reconnaissance will pave the way for these future explorers. And now other nations are joining in the exploration of Mars. The European Mars Express mission has already sent back some extraordinary pictures from orbit, in unprecedented three-dimensional detail and clarity, although some questions have been asked about the bizarre colors in some of the images. Russia, which led the way in planetary exploration in the early 1960s, lost its lead to JPL early on, since when the former space superpower has had an astonishingly grim run of bad luck with its probes to Mars. The future of Russian Martian exploration surely lies in collaborative efforts with its new partners in Western Europe and with the old enemy, NASA.

(*Opposite*) The Viking mission, from the perspective of today's slimmed-down NASA, was almost a no-expense-spared extravagance. But at the time it was seen as very much a cut-down version of the far more ambitious mission planned in the 1960s. Vikings 1 and 2 revolutionized our view of the planet Mars. This image shows Arabia Terra and the great circular bowl of Hellas, photographed by the Viking 2 orbiter.

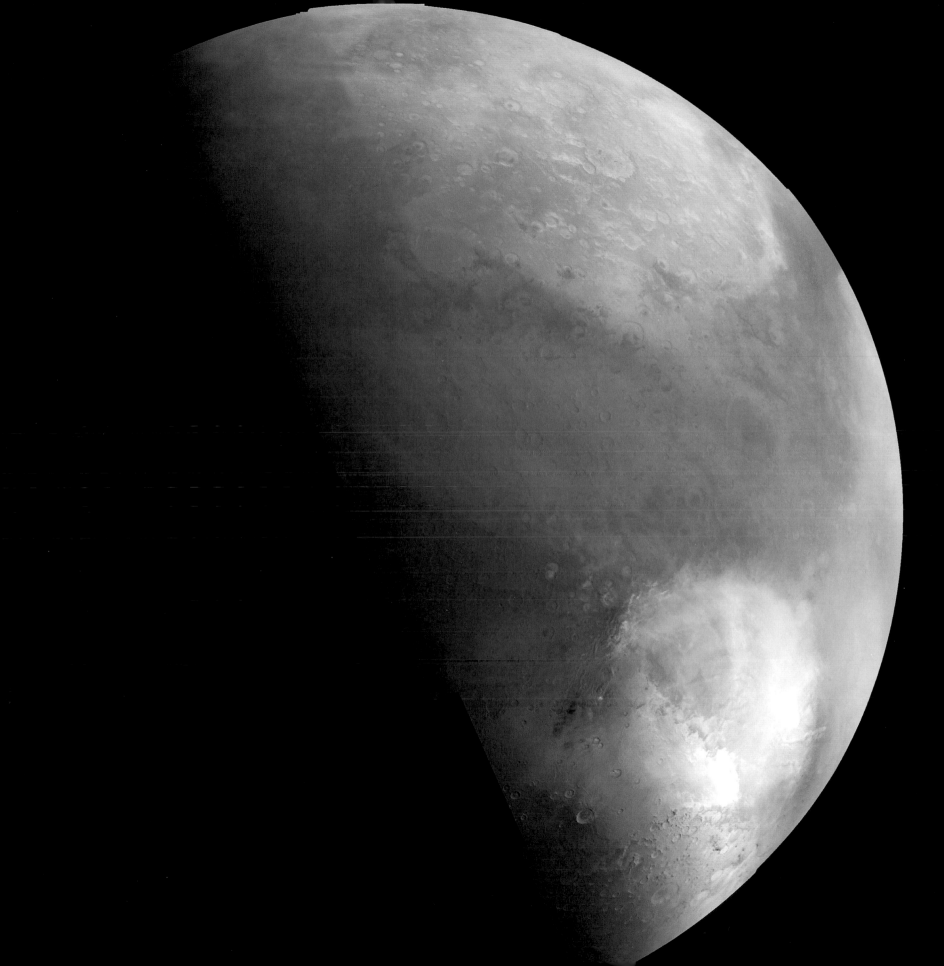

One of the most evocative images from the Viking mission was this beautiful sunset, captured by Viking 1 from its landing site in Chryse Planitia. The picture of the setting Sun – our Sun – seen from the surface of another planet really brought home the fact that Mars was a world in its own right, not just an object in space.

The human invasion of Mars began just forty-four years after the death of Percival Lowell, and although the first brace of probes failed miserably, they deserve mention because jointly they formed humanity's first attempt to visit another planet. Both probes failed to reach orbit after the failure of pumps on their rockets' third-stage thrusters. The very existence of the Soviet Marsniks 1 and 2, launched within a week of each other in October 1960, was a secret. The nascent space race precluded talk of failed missions by the Russians, and the fact that they had tried to send craft to Mars was not known in America until 1962, when the NASA administrator reported intelligence on the failed launches to Congress. Even the scientists involved denied all knowledge of the probes. However, the existence of the Marsniks was confirmed only as recently as 1998, by Vladimir Perminov, the head of the Lavochkin Planetary Probe Design Bureau. The objectives of the Marsnik missions were quite ambitious, and included an investigation of interplanetary space and the return of images from Mars gained from a flyby of the planet. Neither probe was designed to orbit Mars, simply to fly past at several tens of thousands of kilometers an hour in the hope that some useful data could be gleaned on this brief visit. Neither probe succeeded in reaching Earth orbit.

The twin spacecraft each consisted of a cylindrical body about 2 m (6.5 ft) high with two solar panel wings, a 2.3 m (7.5 ft) radio antenna, and a long antenna arm, and the whole probe, including propellants, had a mass of about 650 kg

(1,400 lb). The Marsniks carried a 10 kg (22 lb) science payload consisting of a magnetometer on a boom, a cosmic-ray counter, a plasma-ion trap, a radiometer, a micrometeorite detector, and a spectroreflectometer to study the carbon–hydrogen band, a possible indicator of life on Mars. Back then the possibility of Martian life-forms was considered to be far higher than it is today. These instruments were mounted on the outside of the spacecraft. A photo-television camera in a sealed module in the spacecraft could take pictures through a viewport when a sensor indicated that the daylight Martian surface was in view. It is believed that the camera and the life-detector were removed from the spacecraft a week before launch to save weight.

The next probe to be launched at Mars, Sputnik 22, was notable for almost starting World War III. Launched on 24 October 1962, the Sputnik blew up in Earth orbit, probably because of a catastrophic failure of its upper stage. The explosion caused the spacecraft to break up into several pieces, which apparently remained in Earth orbit for several days. This happened right in the middle of the Cuban Missile Crisis, and when the debris was detected by the American ballistic missile early warning radar in Alaska, the doomed Mars probe was momentarily mistaken for a Soviet nuclear attack. Ten days later another probe, Mars 1, was launched, and although it failed to reach Mars it did manage to get deep into interplanetary space – more than 100 million km (60 million miles) from Earth, and sent back valuable data about the environment beyond low Earth orbit. Although there were problems with the spacecraft's orientation systems, it remained under control until 21 March 1963, when contact with it was lost.

The design of the Mars/Sputnik missions was broadly similar to that of the earlier Marsniks, with life-detection instruments and cameras carried on board to probe the Martian environment. While on the interplanetary cruise, the probe recorded one micrometeorite strike every two minutes at altitudes ranging from 6,000 to 40,000 km (3,750–25,000 miles) from the annual Taurid meteor shower. The radiation belts around the Earth were detected and their magnitude confirmed. Although contact

The Mars 1 probe built by the Soviet Union was the first object to fly to another planet, although by the time it flew by Mars at about half the Earth–Moon distance in June 1963, contact with it had long been lost.

with the probe – and hence the mission – had been lost, Mars 1 did in fact become the first human-built machine to visit another planet; its trajectory suggested that it flew past Mars at a distance of some 197,000 km (122,000 miles) on 19 June 1963. Sputnik 24, launched in November 1962, was the first probe designed to land on Mars. After launch, the probe broke up as the main engine was fired to send it Mars-bound.

So by 1964, there had been five attempts to send robot probes to Mars, all of them by the Soviets and all ending in failure. As with manned spaceflight, the US were beaten off the blocks by the Russians, but soon caught up and surpassed their competitors. It was in this year that the first successful Mars mission, Mariner 4, was launched, on 28 November. Carried aloft from Cape Kennedy (as Cape Canaveral was briefly known) by an Atlas rocket, everything went perfectly, and 228 days later, on 15 July 1965, the probe flew past Mars at an altitude of around 10,000 km (6,000 miles). (Mariner 4's twin, Mariner 3, failed to achieve the correct trajectory for the Mars cruise when it failed to eject its heat shield after launch – another victory to the Mars Ghoul; Mariners 1 and 2 were Venus probes.) Mariner 4 did not enter orbit around Mars, but flew past, sending back a stream of pictures and data.

The Mariner 4 spacecraft 'chassis' consisted of an octagonal magnesium frame, 1.27 m (42 inches) across and 0.46 m (18 inches) high. Four solar panels were attached to the frame, spanning nearly 7 m (23 ft). Under the frame was slung the TV camera, mounted on a scan platform. The Mariner 4 science instruments, in addition to the TV camera, were a magnetometer, a dust detector, a cosmic-ray telescope, a trapped radiation detector, a solar plasma probe, and an ionization chamber/Geiger counter. It was equipped with an onboard computer and a tape recorder onto which data could be stored for playback later. In its basic design elements and concepts, the Mariner 4 probe set the pattern for space probe design for the next four decades.

Although it was known that Mars was probably very cold and very dry, it was not known for how long these conditions had persisted. Mariner 4's twenty-one pictures, sent back pixel by pixel, were black-and-white, crude, and hopelessly fuzzy compared to the brilliantly colored images from today's space probes, but they were a massive advance on anything that had been achieved by telescope. Mariner 4 saw craters – more than seventy of them. It showed that the sharp boundaries

between the light and dark areas were largely illusory; close up, the Martian surface shows nothing of the clear division into bright and dark that was seen from Earth. This probably killed off the vegetation theory which had been put forward to explain the shifting patterns on Mars. This theory was popular right up to the dawn of the space age. In the 1950s, a study suggested that chlorophyll has been detected spectroscopically on Mars, breathing new life into the vegetation hypothesis. It was later found to be a mistake – and anyway, by then it was known that there was almost no free oxygen in the Martian atmosphere. Mariner 4 found that Mars was not the flat world that had been supposed by Lowell and everyone since. Four-thousand-meter (13,000 ft) mountains were glimpsed, although the great volcanoes of Tharsis were not spotted – hardly surprising as less than 1 percent of the Martian surface was photographed. But most importantly, Mariner 4 answered one of the most vexing questions about the planet: how thick was its atmosphere?

Shortly after 2.19 a.m. on 15 July 1964, Mariner 4 passed behind Mars, continuing to send constant-strength radio signals to Earth. By analyzing the way the passage of the radio signals was affected as they traversed the Martian atmosphere, an accurate assessment of its density could, for the first time, be made. The results were incredibly depressing – at least for anyone hoping that Lowell's Mars might yet make a last stand. While before 1964 the best guesstimates for Martian atmospheric thickness hovered around the 80 millibar mark – thin, but not impossibly so as far as life was concerned – Mariner 4 showed that the surface pressure on Mars was between 4 and 7 millibars – a near-vacuum. The thin air was made of carbon dioxide. With these pressures – and temperatures lower than those on the Antarctic plateau – it became clear that Mars could not be the lush world of Lowell, or even the harsh but habitable desert imagined by the astronomers of later decades.

Of course, by 1964 few people actually believed in Martian canal-builders any more (by now telescopes could show details on Mars in great clarity), but the common view of conditions on Mars as similar to that espoused by the British astronomer Walter Maunder in the 1900s, who said that it would be rather like being on top of a 20,000-foot (60,000-m) mountain on Spitzbergen Island, deep in the Arctic. Throughout the twentieth century, that mountain

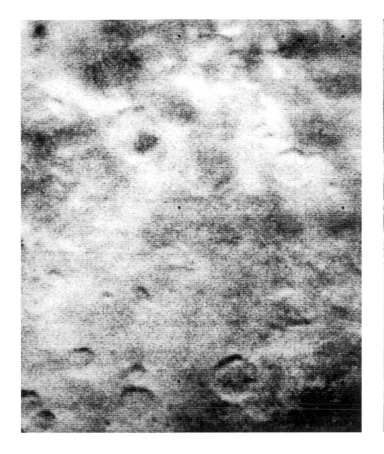 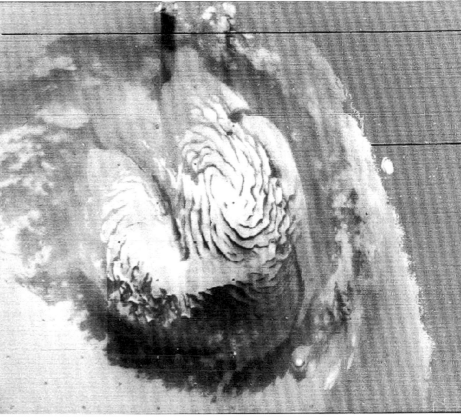

During the 1960s great leaps were made in spacecraft and imaging technology. The image on the left was taken by Mariner 4 in 1965 and shows the craters that dashed the hopes of the romantics waiting to see Lowell's Mars. The image on the right, far crisper, was taken by Mariner 9 in 1972 and shows the dramatic ice-swirls of the northern polar cap.

grew higher and higher. Mariner 4 found that a more accurate analogy would be the top of a mountain five times higher still, located at the south pole. Liquid water would not be stable on the Martian surface – below 6.1 millibars, water can exist only as a solid or gas.

Mars, then, was like the moon – a world of craters, frigid and with shifting sands rather than vegetation. The only difference from the Earth's satellite being that at least there was a trace of air to blow the sand and dust around. Not only were there no canals, but many of the other 'features' on Mars taken for granted by astronomers for

over a hundred years vanished as well. The violet layer was nowhere to be found, for example, nor were the supposed dark bands around the poles. It is no exaggeration to say that the results of Mariner 4 were seen as a 'Great Disappointment' by the astronomical community, and a blow to those who wished to explore this world.

Suddenly an enigmatic, vibrant planet had been revealed as a dry, dead, Moonlike place where, as one astronomer said, 'nothing ever happens'. Mariners 6 and 7, another pair of successful Mars flyby probes, confirmed the Mariner 4 results, with better pictures. Although some features of interest were seen – including bright areas with far fewer craters, and strange areas of what became known as chaotic terrain, the view of Mars as a dead world was reinforced. Impact craters are a sure sign of a surface of great age, as a world that is geologically and meteorologically active will quickly erase these ancient scars. Mars must have been dead since the early days of the solar system.

MARINER 9 AND VIKING: MARS COMES BACK TO LIFE

After the 1969 flyby missions of Mariners 6 and 7, Martian planetary science was moribund. What was historically the most interesting planet in the solar system had been concluded to be an empty world, and attention switched to Venus and plans for the 'grand tour' of the outer solar system. But Mars fought back. The Mars Ghoul had not managed to get the three Mariners in its jaws, but it had struck in a more subtle way. It seems that, purely by chance, the three probes had managed to fly past some of the least characteristic parts of the planet. The areas around the Martian equator photographed by Mariner 6, the Sirenum area examined by Mariner 4 and the cratered southern highlands seen by Mariner 7, were not only relatively 'dull' – i.e. crater-strewn, they were also atypical of the Martian surface as a whole. Although large tracts of Mars *do* look like the Moon, especially from a great distance and through the rudimentary lens of an early-1960s TV camera, far more of Mars is quite unlunar in every respect. Craters, yes, but also towering volcanoes, shifting seas of dunes, whirlwinds and, most importantly, an immense, almost planet-wide network of huge, dried-up courses that look for all the world as though they have been carved by water.

Mariner 9 was the first probe to go into orbit around Mars, and it spent many months redrawing the map of what again became the solar system's second most interesting planet. Arriving on 14 November 1971 after a 161-day flight, Mariner 9 was equipped with the best set of instruments yet aimed at Mars. But again misfortune struck: Mariner 9 arrived at a time when a huge dust storm had enveloped Mars in its entirety, and its surface, even from orbit, was no more than a dull orange ball. For a few days nothing much could be seen, but as the dust cleared the world of Mariner 4 was stripped away. Huge volcanoes, the highest of which corresponded to a feature christened Nix Olympica by telescopic observers, towered to nearly 21 km (13 miles), three times as tall as Everest. About a third of the planet was found to be covered with craters, but the rest was divided into two zones: a rugged, volcanic region of lava flows and few craters (suggesting a relatively young geological age), and a final third consisting of hard-to-explain, alien landscapes, including a

(*Above*) In 1990 it was revealed that the Russians had managed to get two rovers down to the surface of Mars, as part of the Mars 2 and Mars 3 missions of 1971. Each rover was a small, semi-autonomous vehicle attached to the main lander by an umbilical cord, and would have skated along the surface on skis. Both missions failed, although Mars 3 transmitted 20 seconds of data from the surface.

(*Opposite*) The launch of Viking 1, on 20 August 1975, marked the beginning of the end of the first wave of Mars exploration. After the Vikings, it would be another two decades before another successful mission to the Red Planet.

canyon near the equator so big it could swallow the one in Arizona for breakfast, chaotic terrain of hills and ravines, scarps and mesas. And channels; long, meandering valleys that look like dried up Mississippis and Niles. It was the channels that made Mariner 9 the most important mission to Mars to date. The planet that had been written off as no more than a glorified Moon now had a past – and a past to be proud of. A warm, wet past perhaps. A living past.

It was, as Patrick Moore once remarked, 'rather fortunate that plans for two more Mars probes, Mariners 8 and 9, were so well advanced that there was no thought of canceling them'. Mariner 8 was lost at launch, but Mariner 9 saved Mars exploration by the skin of its teeth.

Meanwhile, the Soviets had not given up on 'their' Red Planet. Mars 2 and Mars 3 were also launched in 1971, and both made it into orbit around their destination, although they returned only a small fraction of the data sent back by Mariner 9. These Soviet probes carried large and complex landers. These impressive machines, weighing more than a third of a tonne apiece, were each equipped with two TV cameras, a mass spectrometer to study the atmospheric composition, a weather station, and devices to measure mechanical and chemical properties of the surface, plus a mechanical scoop to search for organic materials and 'signs of life'. They also contained pennants with Soviet insignia. The landing capsule was sterilized before launch to prevent contamination of the Martian environment.

The Mars 2 and 3 landers each carried a small, 4.5 kg (10 lb) walking robot called PROP-M. The existence of these rovers was kept something of a secret until 1990, and all official descriptions of the Mars 2 and 3 probes omitted any mention of a rover. The small vehicles were attached to the main landers by tethers, giving a range of some 15 m (50 ft) from the lander base-station. Unlike all later rover designs, the Russian rovers had motorized skis rather than wheels. They were equipped with primitive artificiaal intelligence: sensors could detect obstacles in the rover's path, and onboard electronics could interpret this information and redirect the machine as necessary. The rovers carried two scientific instruments: a dynamic penetrometer and a densitometer, to measure the strength and density of the soil. A year previously, the pram-like Lunokhod 1 had successfully become the first mobile spacecraft to land on another world, in this case the Moon. The Russians very much hoped to repeat this triumph on Mars.

Mars 2's descent module successfully deployed its parachute but crashed, although it did hit Mars, making it the first object from Earth to make contact with the planet. It seems that the lander ejected from Mars 3 made it to the surface in good health. It even started transmitting a picture, although no detail was seen and transmission stopped after 20 seconds. The rover was not activated. It is not known why; perhaps the lander was damaged by a collision with a 'terrain anomaly' (i.e. a rock). More likely, a dust storm raging at the time – probably the most severe ever seen on Mars – may have interfered with the radio transmitter, or simply blown the spacecraft over or flipped the collapsed parachute over the radio transmitter.

While the Soviets were the first to get a spacecraft onto the surface of Mars, they were trumped yet again by the Americans in the shape of the magnificent Viking mission of 1976. Between them, the two Viking orbiters and their landers performed flawlessly, transforming once again humanity's view of Mars. True, there was to be no sea change as with Mariner 4, and then back again with Mariner 9. The Vikings did not change what we know about Mars for ever. But they did bring Mars to life, and brought an alien world into our living rooms.

THE VIKING MISSION

The Viking program was a cut-price replacement for the Voyager Mars program after the latter was killed off by the deep NASA budget cuts in 1967. As early as 1960, JPL had envisaged Voyager landers and rovers as ambitious follow-ons to the Mariner flyby/orbiter probes. Congress approved Viking's development in 1968, with an eye on launching two orbiter/lander combinations to Mars in 1973. Budget trimming delayed the launches to 1975.

Back in the 1970s, it was thought madness to spend a lot of money designing a complicated spacecraft and sending only one on its mission to another planet. Much of the cost was in the development phase, and the built-in redundancy of sending a pair of probes was a feature of NASA planning until the cost-cutting really took hold. So there were two Vikings (just as there were two Mariner '9's), twin missions designed to carry out identical tasks on two completely different regions of the planet. The twin landers were about 3 m (10 ft) across on their splayed legs and weighed more than half a tonne each. Both machines were sterilized in a pressure cooker for more than 40 hours, although it is doubtful that this removed all the microorganisms that would have infested them.

The two-tonne orbiters, basically enlarged and modified versions of the old Mariner platform, were extraordinary machines. Viking 1 carried on sending back data until August

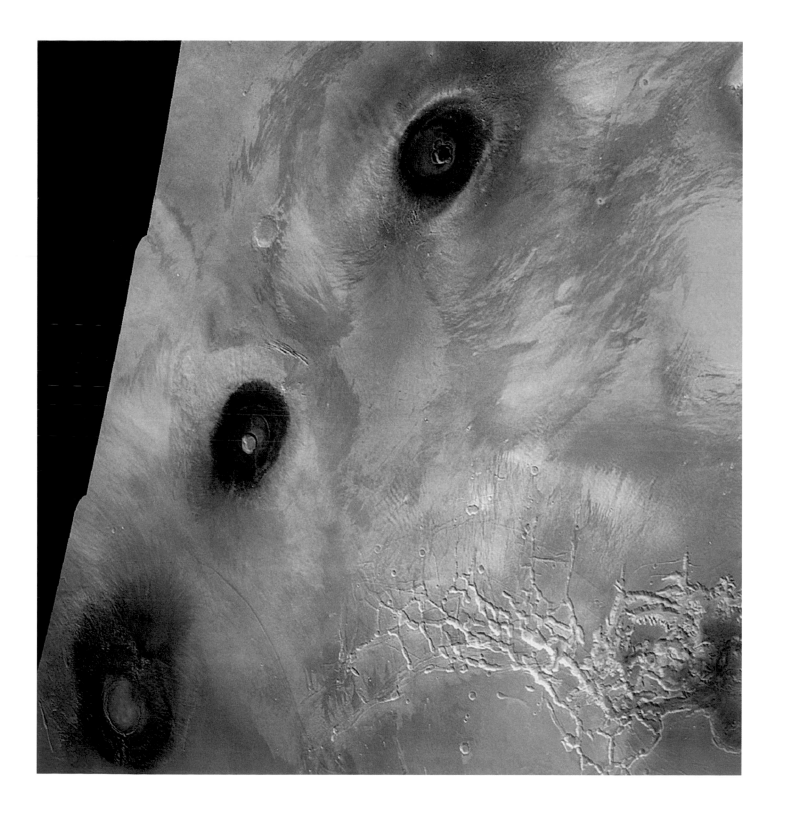

The complexity of the Viking mission was unprecedented. The fact that both probes survived not only the long voyage to Mars but also a perilous descent to the surface, and then performed way beyond their nominal mission, is a testament to the quality of their design and engineering.

Martian moon. The Mars jinx struck again five years later, this time at NASA. The billion-dollar Mars Observer Mission ended in August 1993, when (it is presumed) a fuel line burst just three days before it would have entered orbit around Mars – a terrible blow for JPL. Observer was supposed to map Mars from orbit and try to determine the makeup of its rocks.

THE GHOUL VANQUISHED

Twenty years after the Vikings began their extraordinary missions, the new invasion of Mars began. Since 1996, a veritable armada of robots have been flung at the Red Planet, with varying degrees of success. Enough of them have made it to transform our ideas about Mars, and some of the data they have returned has confirmed the suspicion and hope of many scientists that the cold, dry world of the Mariners was perhaps over-pessimistic. Detailed photography from orbit has revealed the Martian surface in detail that would have seemed improbable to the designers of Mariner 4. The photographs sent back by the camera on board Mars Global Surveyor (MGS) form not just a scientific data set that will probably be mined for decades, but also a new and beautiful take on what is turning out to be a very photogenic planet.

Under the skillful and PR-mindful directorship of Mike Malin, the man subcontracted to run the Mars Orbital Camera (MOC) on board MGS, Mars has been reinvented as a world of frost-covered dunes, enigmatic ice-pits, heart-shaped craters (an image that was released on Valentine's Day), and sweeping canyons that surpass anything in Arizona. With images of this quality, Mars has transcended science and become art, and to a degree its surface has now become public property. Malin's pictures are now familiar not just from the science pages of newspapers and the occasional TV reports when something truly startling turns up, but from posters and schoolroom walls all over the world. Some features of the Martian landscape are at last becoming familiar in a way that even the Moon never did. Most people with even a passing interest in astronomy are now familiar with the monster canyon named after the Mariners, and of Mount Olympus, the Nirgal Valley and the glittering polar ice. Thanks to Mars Global Surveyor, the planet has become a place.

For such a successful mission, it is surprising to think that MGS was something of a last-minute fix, in NASA's words a 'rapid, low-cost recovery of the Mars Observer mission objectives'. In fact, MGS was part of the Discovery Program, instigated by NASA administrator Dan Goldin as part of his 'faster, better, cheaper' philosophy. The idea was to build cheap spacecraft out of spare parts and launch them as quickly as possible, adopting a scattergun approach and hoping that at least some of them worked. The new philosophy was

supposed to be an antidote to the old 'Rolls-Royce' missions where lots of instruments were piled onto a large and costly platform, at a price that allowed only single launches, or twin launches at most, every few years. Less than half the price as its lost predecessor, MGS was a stripped-down version of Observer which has nevertheless done sterling work after a shaky start. A lack of fuel meant that to slow down enough to reach the required orbit around Mars, the spacecraft performed a risky maneuver called aerobraking – dipping into the upper reaches of the Martian atmosphere to use air friction

Mars Global Surveyor has been orbiting the planet since 1998. After a shaky start, when it was found that one of the solar panels had not locked into position correctly, the spacecraft has performed flawlessly.

to slow it down. On arrival at Mars, one of the solar panels was found to have jammed in the wrong position, making the aerobraking procedure potentially hazardous and delaying mapping of the planet's surface for over a year.

After all that, though, MGS has proved to be an astounding success. The Mars Orbital Camera, the Mars Orbiter Laser Altimeter and the Thermal Emission Spectrometer have prodded and probed the planet from orbit, using photons rather than a rock hammer. MGS is still going strong, and its work is now being supplemented by Mars Odyssey, which reached the planet in January 2001.

The same year that Mars Global Surveyor arrived, another spacecraft landed on Mars. Mars Pathfinder proved to be a big hit – not so much for breaking any new scientific ground but for two other things. Pathfinder was a proving ground for a new technology – a small, semi-autonomous lightweight electric rover (in this case called Sojourner) and a wholly novel landing system, using a packet of airbags to protect the descending probe. It was also notable for its cost – around $265 million, around a tenth, in real terms, of what the twin Vikings had soaked up. Pathfinder was supposed to pave the way for a whole new generation of cheap rovers, but in the end (as is often the case with NASA projects) it became a one-off. It

proved that the airbag system worked, much to the chagrin of some engineers who do not trust the technology to this day (airbags were also used, equally successfully, on the Mars Rovers). Pathfinder proved that you could do Mars on the cheap, but this too was a mixed blessing. 'That probe', one NASA employee once told me, 'was the worst thing that ever happened.'

But Pathfinder duly found itself in Ares Valles on 4 July 1996. Pathfinder was notable for being the first space probe to become famous via the Internet. Several hundred million

(*Above*) Mars Pathfinder in 1997 was the first space mission to hit the Internet. For a while, the JPL Pathfinder site was the most popular on line, with hundreds of millions of hits a day. The small rover vehicle, Sojourner, captured the public's imagination in a way that no other robot had done before. It is shown here examining a rock that was nicknamed Yogi.

(*Opposite*) Mars Pathfinder was so named because it was a testbed for a host of new technologies, most notably the cushioned airbag landing system and the wheeled Sojourner rover. Tested extensively in the world's largest vacuum chamber at NASA's Lewis Research Center, Pathfinder's airbag system was refined until it was as safe and reliable as a retro-rocket system weighing – and costing – far more. Airbags have been used since, both on NASA and ESA missions, and despite their complexity have had a good success rate.

'hits' were recorded on the Pathfinder website during the three months for which it remained in operation. Pathfinder studied a host of interesting rocks which were mostly given the names of cartoon characters – 'Yogi', 'Stimpy', and so on. In the old days, these things were given names in Latin.

Pathfinder saw that the valley it landed in was a far-from-random jumble of boulders – the rocks appeared to have been laid down in some sort of directional flow, possibly a fast-moving river or outwash flood. It detected silicate minerals in the rocks, but if truth be told the science results from Pathfinder and its rover were limited. Tight bandwidth from a tiny transmitter restricted the amount of data that could be transmitted back to Earth, and in fact only four full panoramic shots were taken during the whole mission. It has been pointed out that despite the landing site being awash with dust devils, not one was seen in action by the lander when it was active. If any little green men had been in the vicinity of Pathfinder, they could probably have wandered merrily around the probe with little chance of being detected. But to be fair, Pathfinder was very much a technology demonstrator – and as such it was a stunning success.

The ambitious Russian Mars 96 mission came to grief over the mountains of Chile and Bolivia after the failure of its launch system. A real loss to Mars science, this ambitious probe was designed to send an orbiter, two small autonomous landers, and two surface penetrators to Mars. Several of the instruments on Mars 96 were 'rescued' as concepts, and identical versions were flown on a stripped-down version of the probe, built by the European Space Agency and known as Mars Express (see Chapter 14).

MARS ODYSSEY: A NEW CENTURY DAWNS

Since its arrival at Mars on 24 October 2001, Mars Odyssey has lived somewhat in the shadow of its more glamorous sibling, Mars Global Surveyor, thanks to the latter probe's star instrument – the Mars Orbital Camera. Odyssey is the surviving part of the Mars Surveyor 2001 project, which originally consisted of two separate missions, the Mars Surveyor 2001 orbiter and the Mars Surveyor 2001 lander. The lander was cancelled as part of the reorganization of the Mars exploration program. The orbiter was renamed Odyssey and given a nominal three-year mission to orbit Mars and undertake a comprehensive mineralogical and geological survey of the planet's surface using a pair of instruments, the Thermal Imaging System (Themis), which maps the Martian surface in visible and infrared light, and the Gamma Ray Spectrometer, which maps the elements present on the surface. Odyssey has also played a vital role in acting as a relay station for the two Mars rovers which landed in early 2004. It was Odyssey that was assigned the task of listening out for Beagle after its arrival on Christmas Day 2003, and which failed to produce the good news that Europe and Britain were looking for.

But Odyssey's major achievement to date was almost certainly the discovery, in early 2002, that large quantities of water ice lay just below the surface at both high and lower latitudes on Mars. Using its Gamma Ray Spectrometer, under the command of William Boynton, Odyssey found massive quantities of hydrogen. To the spectrometrists, 'hydrogen' means water – not absolute proof, but strong circumstantial evidence at least. The 'water ice' stretches from the poles well into the Martian 'temperate' zone. The total amount of water was calculated to be enough to fill Lake Michigan twice over. So much water ice was detected that Boynton preferred to call it 'dirty ice' rather than ice mixed with dirt. In fact, over much of Mars the first two meters (6 ft) or so of regolith may consist of some 50 percent of ice by volume. This was not, of course, the first time that water ice had been discovered on Mars – there is a lot of the stuff visible at the poles – buts its ubiquity and abundance came as a real surprise, and contributed greatly to the 'rehabilitation' of Mars after the Great Disappointment of the era of Mariners 4, 6, and 7. With Odyssey's confirmation that much of the Martian regolith is, in fact, permafrost, the stage was set for the twenty-first-century follow-the-water assault on the Red Planet by the probe's successors.

THE GHOUL STRIKES AGAIN: THE DISASTERS OF THE LATE 1990S

To date, the majority of Mars missions have failed, and the most depressing string of disasters came in the final years of the twentieth century, as no fewer than five missions came to grief on the way to, at, or on the way down to the surface of Mars. It probably isn't appreciated by the public just how disheartening the loss of a spacecraft is. These aren't manned probes, after all. But the reality of a failed mission is something truly soul-destroying. Hundreds of men and women have all sunk large chunks of their lives into the design and operation of the spacecraft and its scientific instruments. The loss of a single instrument – say an orbiting camera – is a professional tragedy not only for the people who designed and built it, but also for the academics and investigation teams who were banking on using its data for years to come. Grants will have been obtained, research assistants hired, all on the back of a successful space mission. Should it fail, it is no exaggeration to say that people's lives can be ruined.

There were tears at JPL towards the end of 1999 when not one but two major Mars missions were lost in the space of just three months. The agony was compounded when it was realized that the loss of Mars Climate Orbiter (MCO), a Martian weather station and part of the expanded Mars Surveyor program (the expansion was partly the result of the renewed interest in Mars following the discovery of possible fossils in the meteorite ALH 84001), was down to a mix-up over imperial and metric measurements sent to the probe. The spacecraft reached Mars on 23 September. It then passed behind the planet, entering a communication blackout for what was supposed to be 21 minutes, but contact was never reestablished. The resulting inquiry concluded that some of the commands were sent to the spacecraft in imperial units, instead of the metric units that the onboard computer was expecting. As a result, MCO was sent hurtling into the Martian atmosphere and burnt up, rather than entering orbit at its intended 145 km (90-mile) altitude. The error was a source of huge and very public embarrassment for NASA.

The loss was also seen as a symptom of the sort of cost-cutting and skin-of-the-teeth attitude that typified 'faster, better, cheaper'. Less embarrassing, as there was no obvious human error involved, was the loss of Mars Polar Lander (MPL) together with its complement of 'Deep Space' penetrator probes that constituted a separate mission. The last telemetry from MPL was received on 3 December 1999, just before the descent sequence was scheduled to have begun. Nothing more was heard from the probe, which was supposed to parachute down to the layered terrain near the south polar region and examine the near-surface environment. There was no sign either from the two penetrator probes, which were supposed to smash into the Martian surface and send back signals from underground, hopefully reporting the presence of permafrost.

The loss of MPL was another crippling blow to NASA. While the manned spaceflight program was floundering, at least JPL could be claiming to keep the beacon alight for the Agency with its interplanetary probes. NASA had shown as early as 1976 that it could get not just one but two large, heavy, and complex Mars landers down onto the surface in one piece. How could it have failed with technology two decades' more advanced?

As with all spacecraft failures, an enquiry board was set up, this one headed by JPL veteran John Casani. Reporting in January 2000, the Casani Report was released to a packed auditorium at JPL. Casani himself was not there, and neither was Dan Goldin, the architect of the philosophy that many JPL engineers muttered was the root cause of the mess they were now in. Instead, the report was presented by NASA's Associate Administrator for Space Science, Ed Weiler, and Tom Young, head of the Mars Program Independent Assessment Team. It was a grim press conference. According to reporters who were there, both men looked as though they would far rather have been somewhere else.

The report's conclusion was that the computer on board the lander thought the jolt of the landing legs' deployment was touchdown – causing it to shut down the probe's descent engines immediately they were deployed during the last part of the descent. This was unfortunate as the lander would have still been 40 m (130 ft) above the surface, and would then have plummeted, unrestrained, into Mars at 100 kph (60 mph), smashing itself to bits. Furthermore, the report concluded that the Deep Space 2 mission, in which the two probes Scott and Amundsen were designed to fire into Mars at 600 kph (375 mph), was so poorly conceived and under-tested that it should never have been launched.

The report was, and could not have been, definitive. No telemetry was received from the probe during descent (this was as planned – to save weight there was no external transmitter), so the investigation team had to rely entirely on post-mission ground testing of certain key components to see which one or ones were most likely to have been the cause of failure. Four tests by Lockheed Martin, the prime contractor, found a glitch in the software that analyzed signals from the landing gear. It was known that spurious signals – say those from

FASTER — UNREALISTIC LAUNCH SCHEDULE

CHEAPER — SEVERE NASA BUDGET CUTBACKS

BITTER — MARS POLAR LANDER FAILS

the legs jolting as they were unfurled – could confuse the descent program, and the software should have taken account of that. But it did not. This design flaw was not caught before launch since a series of tests used sensors that were wired incorrectly. Therefore the problem was not detected, and the spacecraft was doomed.

It was now time to reconsider and regroup. Once again, as with the MCO failure, the root cause of the disaster

(Opposite) Mars Polar Lander was an ambitious mission to explore the Martian high latitudes; its failure was down to an error in the controlling governing its descent systems.

(Above) 'Faster, better, cheaper.' It may sound good on paper, but many in NASA and outside the agency have criticized the policy of cut-price space missions, arguing that exploring the planets simply cannot be done on the cheap.

was essentially managerial, not technological. Any spacecraft will have potentially fatal design flaws; software is always fallible, something will always be overlooked. This has been true since the 1960s. But the trick, say NASA's critics, is to have management systems in place that can spot these glitches before probes are launched. The metric/imperial confusion that led to the demise of the Climate Orbiter was caused, fundamentally, not just because a young computer programmer at Lockheed Martin goofed, but because of a failure of communication between NASA and its prime contractor. Getting the human stuff right is as important as making sure that the widgets work.

The causes of the MPL crash were peripheral, in retrospect, to the consequences. Ed Weiler said that the whole Mars exploration plan would have to be refocused. The demise of Polar Lander marked the beginning of the 'search for past and/or current life on Mars', using a follow-the-water paradigm that is now well entrenched at JPL and NASA HQ. Weiler spelt out the need for a more in-depth investigation of Mars that went beyond one-way probes: 'we may need to bring drills to Mars at some point', he said.

The Japanese learned the hard way about the Mars Ghoul on their first attempt at Mars exploration. Nozomi ('Hope') was a half-tonne probe launched in July 1998 that was meant to study Mars's upper atmosphere. Sadly, things started to go wrong very early on. Nozomi was supposed to arrive at Mars in October 1999, but a propulsion failure during an Earth 'slingshot' maneuver in December 1998 meant it could not reach Mars. So a new trajectory was planned which incorporated two further Earth flybys, and an arrival at Mars more than four years late – in December 2003. In April 2002, a large solar flare damaged Nozomi's guidance systems, and despite heroic efforts to save the probe, as it approached Mars it became clear that it was never going to achieve orbit. In fact, there was a small but real risk that it would actually collide with Mars. On 9 December 2003 the mission was abandoned, and Nozomi was 'retargeted' to fly by Mars. 'It is a shame to lose Nozomi as a Mars explorer', said Louis Friedman, Executive Director of the Planetary Society in an article on the Society's website. 'But no one should lose heart. This was Japan's first mission to a planet. They conceived and operated it brilliantly, but space is unforgiving as we have so frequently learned.'

There is, of course, no such thing as a Mars ghoul or jinx. The Red Planet claims its victims because of human error, design flaws, and sheer bad luck. The rather sorry history of Martian exploration – more than half the missions to date have failed – reflects the difficulty of performing deep-space exploration. After the successes of the 1970s, Mars exploration hit the buffers with a series of failures. Now, it is back on track again with the sterling work being performed by the Mars Exploration Rovers and Mars Express.

Chapter 8
Faces and Pyramids

'Artificiality is one hypothesis'
Mark Carlotto (2003)

Type the words 'Mars' and 'face' into Google, and you will be told that your search has been successful, at time of writing to the tune of around 3.5 million hits. The number appears to be growing every day. Of course, things Martian cannot hope to compete with the Internet's star topic, sex, but

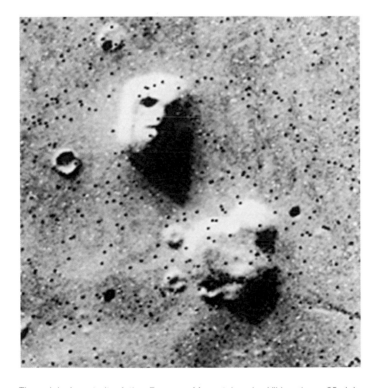

The original portrait of the Face on Mars, taken by Viking 1 on 25 July 1976. The black dots are noise – artifacts of the imaging process. But one dot has fallen fortuitously where the right 'nostril' of the face would be, adding to the illusion.

even so, if you type 'Pamela Anderson' into the same engine the result is a rather less impressive 453,000 (although the charms of Ms Anderson have no doubt become rather passé since she was declared Queen of the Internet some years ago).

The Face on Mars phenomenon is one of the most fascinating in the history of our relationship with this planet. At first reading, it resembles closely the 'canalist' debates of a century ago. After all, those who believe there is a vast artificial stone monument on the Martian surface today and those who believed that there were thousands of kilometers of artificial canals on the Martian surface back in the 1890s were essentially drawing the same conclusions. Most essentially, of course, both phenomena share the same property of being just plain wrong.

But in other ways, the canalists and the face brigade have little in common. The observations made by Schiaparelli and Lowell which gave rise to the widespread belief, in North America at least, that there was an extant technological civilization on the planet Mars were serious scientific observations carried out with some of the best scientific instruments of the day. The results were debated in the pages of *Nature* and other august journals. Belief in Martians flew in the face of no known scientific facts about the Red Planet. In 1900 it was thought that Mars was far more conducive to life, our kind of life, than we know it to be today. Lowell and his contemporaries had no reason to disbelieve the evidence of their own eyes that there were artifacts on Mars. The trouble, as it turned out, was that their own eyes, and brains, were less than honest witnesses. Humans look for order in random data – we turn white noise into tunes and see animals floating across the sky in the clouds.

METHINKS IT IS LIKE A FACE

Few of the people who claim there is a face (and pyramids, a city even) on Mars have ever observed the planet directly with a telescope. Some have, of course, and some have spent a lot of time analyzing the photographs sent back by the space probes. Of course, no one would never claim that the face is visible in a telescope, but this does illustrate that the face controversy is very much a social phenomenon, driven by the information-engine that is the Internet, as opposed to a scientific phenomenon, underpinned by direct observation. The story of Cydonia is a fascinating one in its own right and deserves a chapter in any biography of Mars.

It all began on 25 July 1976, when the Viking 1 orbiter was taking pictures from orbit of the Cydonia plateau, a rugged, windswept terrain of blocky mesas and flat-topped hills a few hundred kilometers south of the equator. Cydonia was being examined as a possible landing site for Viking 2. On that day, from an altitude of about 1900 km (1200 miles) the spacecraft took a series of images showing the rocky buttes and mesas. Among the hills was one that, at first glance, looked oddly familiar. JPL, keen to maintain interest in the mission, put out a press release which drew attention to the image thus:

> The picture shows eroded mesa-like landforms. The huge rock formation in the center, which resembles a human head, is formed by shadows giving the illusion of eyes, nose, and mouth. The feature is 1.5 kilometers (one mile) across, with the Sun angle at approximately 20 degrees.

This was, in retrospect, one of the most ill-advised press releases in NASA's history. It led to decades of controversy, accusations of cover-ups and conspiracies, and even a successful attempt to grab valuable hours of observing time on a later spacecraft. Looking back, one could argue that JPL should have known what it was getting itself into. Although the press release made it clear that any resemblance to a 'face' was purely a trick of the light, the cat was out of the bag. The giveaway word 'illusion' wasn't going to fool anyone.

Interestingly, not a lot happened straight away. In the 1970s, information spread slowly, by book and journal, word of mouth, and in the media. No one took a lot of notice of the Face on Mars, and the absence of an effective Internet meant that rumors took a long time to spread. But a few years later, after computer enhancement sharpened up the images, tales began circulating that NASA had discovered signs of an ancient civilization on the Cydonia plateau. There was not just a 'face': – the Viking photo seemed to show a host of other 'anomalous features' as well, including 'pyramids' and 'archaeological remains' arranged in precise – and significant – geometrical patterns. Now the 'face' had become a 'Sphinx', and people were drawing analogies between Cydonia and Giza in Egypt, site of the great pyramid of Cheops. By the 1990s, more than fifteen years after the photograph was released, a whole community of Face-watchers took up residence on-line.

A glance at some of the half a million Face-on-Mars websites gives a flavor of the conspiracy. One Montana-based site (www.mt.net/~watcher/mars.html) sums it up:

> This image was initially dismissed as a trick of the light by NASA. Other images were discovered which had been taken by the Viking probe at a different sun angle ... and the face was still there. Also present on these photographs were additional anomalies perhaps even more impressive than the face itself. One of the greatest of these, a five sided pyramid nearly two miles [3 km] wide arranged exactly 1/360th of the Martian polar diameter from the face. This pyramid was also pointing to the face. There was also discovered what appeared to be a city complex so oriented that all three structures arranged to form a perfect equilateral triangle.

There are many versions of the Face story, but they all maintain that the Cydonia plateau is home to artifacts. A typical mythology goes something like this: at some unspecified time in the past, a race of alien beings from who-knows-where constructed a series of artifacts on Mars. These beings also visited the Earth, where they constructed the Egyptian pyramids, along with various other objects. A 'city' now exists on Mars, on the Cydonia plateau, that bears an uncanny resemblance to the Giza area of Egypt. The 'Face-builders' may still be around, we do not know. NASA discovered these facts 25 years ago and has been trying to cover them up ever since, terrified that the public will discover the 'truth'.

The Face on Mars is now so well known that it has crept into the popular consciousness, in much the same way

The 'Cliff'

The 'Face'

The 'City'

The 'City Square'

The 'Tholus'

The 'D&M Pyramid'

that the 'fact' that the Apollo Moon missions were faked has become received wisdom in some quarters. Spread by the Internet, and cropping up in popular fiction like the *X-Files*, the Face has taken on all the attributes of a *bona fide* modern myth. The Face could even be said to be part of what may be a minor modern religion, a religion, so its adherents would say, based on 'science' rather than 'superstition'. Belief in the Face on Mars tends to go along with a series of related creeds – a general New Age tendency, an interest in the ancient structures of Europe such as Stonehenge, a fervent acceptance of the existence of flying saucers, a belief that the patterns seen in cereal fields every

The 'anomalies' of Cydonia. The blocky mesas and eroded hills have been interpreted by wishful thinkers as a carved sphinx-like face, a manufactured pyramid, and the remains of a city. The image shows an area about 100 km (62 miles) across.

English summer are not just the work of pranksters and bored farmers, and a general suspicion that there is Something that we are Not Being Told. And, most worrisome of all, 'They' are in control.

Face-believers back up their claims with reams of evidence – photographs, calculations, even some quite impressive docu-

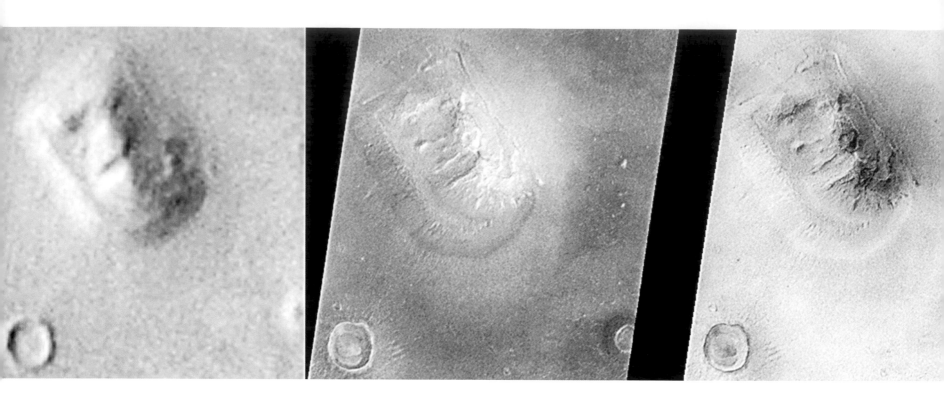

ments. Belief in the Face on Mars is extremely similar, it seems, to belief in crop circles; even when people confess that they went out with some friends one night with some string and a garden mower, the believers still insist that 'most' of the crop circles are real – just as when NASA obligingly takes close-up pictures showing that the Martian artifacts are just natural eroded hills, most people still insist there is a cover-up. (In one book, *Dark Moon* (1999), authors David Percy and Mary Bennet manage the impressive feat of uniting the crop circles, the Face on Mars, Ancient Egypt, and the mystery of Stonehenge! The only wonder in this book is that neither Elvis Presley nor JFK make an appearance.)

Viking's grainy photographs, which could be interpreted this way or that, were never going to be the end of the matter. To most casual observers, including this one, that hill in Cydonia *does* look a bit like a face. A second oppor-

tunity to get a look at the Cydonia region came in 1998, courtesy of the Mars Global Surveyor spacecraft, equipped with its far keener-eyed Mars Orbital Camera. On 5 April, JPL and Mike Malin obliged the public by snapping the area that had caused so much fuss as MGS flew over Cydonia. The resulting high-quality images, since added to on subsequent passes over the region, show once and for all that the 'face' is no more than a trick of the light – just as JPL insisted in those innocent, pre-*X-Files* days of 1976. There are no sphinxes, no pyramids, just a low, stubby hill.

THE D&M PYRAMID, WORMS, AND WHALES

The first of the 'Cydonian anomalies' to be recognized was the Face. After being more or less forgotten, the Viking image showing this vaguely humanoid feature was reportedly 'rediscovered' by two computer scientists working at the Goddard Space Flight Center, Vincent DiPietro and Gregory Molenaar. They also noted the presence of a large pyramidal hill, nearly 3 km long and 1 km high (about 2 miles by 1 mile), to the south of the 'city' (a group of squarish, steep-sided features scattered over the Cydonia plain near the Face). The hill was subsequently named the 'D&M pyramid' and much was made of its geometry, and the fact that its various

(*Above*) Fading away: the Face on Mars, Shown at the left in a Viking image, slowly resolves itself into a hill in the two Mars Orbital Camera images, center and right.

(*Opposite*) On closer inspection (in this case by the Mars Orbital Camera aboard Mars Global Surveyor), the 'face' is revealed for what it is: a knobbly, eroded hill about 2 km long (over a mile) and a few hundred metres high, that could be climbed easily in a couple of hours.

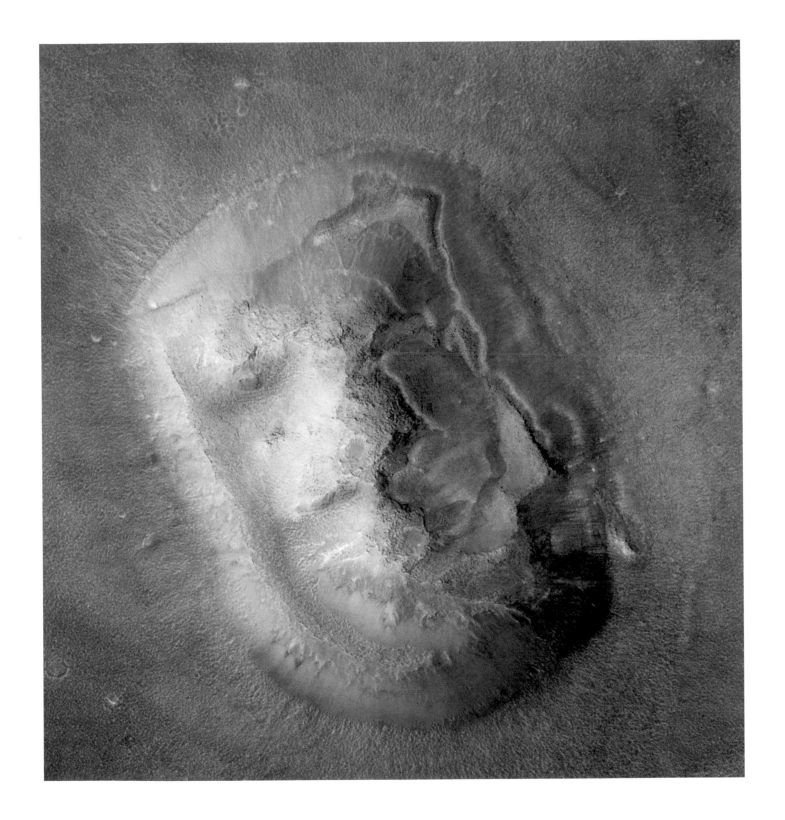

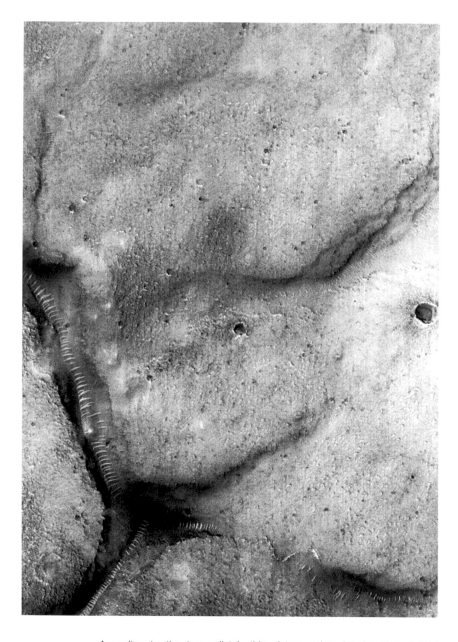

According to the 'anomalists', this picture, taken by the Mars Orbital Camera of a region of Acidalia Planitia, shows three large, ribbed glass-covered tunnels about 100 m (330 ft) wide converging in a valley or gorge. This is no more than a trick of the light – or rather a trick of the brain. What we are seeing are lines of parallel sand dunes lying in troughs. They look superficially like the ribs on a tunnel structure – trying in vain to make any sense from the other visual cues in the image, this is what our eyes tell us they are.

facets appeared to point to other 'anomalous' objects in Cydonia.

Perhaps the strangest 'anomaly' to be spotted on Mars are the giant 'sandworms' or 'glass tunnels'. A picture taken by Mars Global Surveyor in August 1998 shows a pair of intersecting valleys, or gorges, along the floors of which are dozens of slender ribbed structures. Yes, they *do* look like supporting ribs on a gigantic network of glass tunnels – or maybe 'giant worms', as one or two websites claimed. In fact, as with so much on the Martian surface, our eyes are playing tricks with us. These are not ribs, but sand dunes ranged in parallel along the valley floor. At the poles, the alien carbon dioxide landscapes give rise to other bizarre landforms which our brains struggle to make sense of. Deposits of dust on the south polar ice look like 'trees' or bushes'. One engaging book, *The Sand Whales of Mars* interprets a host of features on the Martian surface as, variously, gigantic mating animals, skeletons, and 'fungi'.

THE FACE ON MARS EXPLAINED

There are two explanations for the Face on Mars. The charitable one is that the phenomenon is a classic example of pareidolia, a type of illusion or misperception in which a vague or obscure stimulus is perceived as something clear and distinct. For example, in the discolorations of a burnt tortilla one sees the face of Jesus Christ, or one sees the image of Mother Theresa in a cinnamon bun (yes, this has really been reported) or the face of a man in the Moon. Pareidolia may explain many so-called mystery phenomena, from UFO sightings to Bigfoot, the Loch Ness Monster to ghosts. It explains numerous religious apparitions and visions. And it explains why some people see a human form in a photograph of the Cydonia region of Mars. Some uncharitable souls, however, accuse the people promoting the idea of a Face on Mars, through their websites and books, of fraud.

Probably the best-known of the people associated with the Face on Mars controversy is Richard Hoagland. Based in New Mexico, Hoagland has forged a career promoting his ideas, in books, on the Internet, on TV, and on radio, about the Cydonia plateau. His book, *Monuments of Mars, a City on the Edge of Forever* (1996), was something of a best-

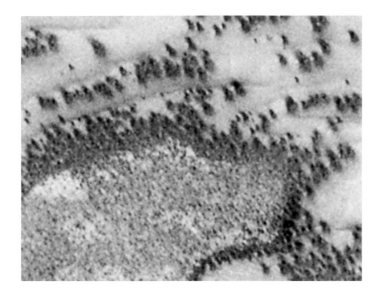

These are sand dunes, just north of the South Polar cap, defrosting in the spring sunshine – covering an area shown of 1.8 km (1.1 miles across). As they warm, carbon dioxide frost sublimes off the surface, exposing dark material below. That is the reality. But these spots do look superficially like shrubs, or pine trees, leading the anomalists to claim that we have already found life on Mars.

seller. At time of writing, a follow-up is in the pipeline. A few days before he and I met, by several scientists at the 2003 Pasadena Mars meeting had warned me what to expect; most of their descriptions of Hoagland are unprintable. He is probably the single most unpopular figure among scientists who study Mars.

So when we did meet, I was half-expecting a gibbering, demented lunatic. In fact, Hoagland in the flesh is quite sane, charming, and articulate. Immaculately turned out, not a hair out of place, he comes across as a Western Gentleman. He was accompanied by his rather glamorous partner, Robyn (a former actress or model, as far as I could make out). I met the Hoaglands in Flagstaff during the opposition of 2003. I wanted to interview him, and he suggested the dome of Percival Lowell's Clark telescope would be the ideal venue, something which caused some consternation among the Observatory's PR department once they realized who Hoagland was. Hoagland had brought along a small TV crew with him; he wanted to film me interviewing him for a documentary he was making.

If you spend any time at all with Richard Hoagland, you start to slip into a strange parallel universe. Hoagland's ideas stretch much further than to Mars. What he believes in is not so much a conspiracy theory as an all-embracing conviction that the entire world – certainly the world of space exploration – is something of a con trick. Before meeting him I spent some time at his website. All the Cydonia stuff was there, together with some new material, a series of images from the NASA Pathfinder website. Hoagland's team had highlighted some of these images to show some extraordinary things which, the website claimed, the JPL robot had discovered on Mars. These included artifacts such as tanks and guns. One picture showed the Pathfinder lander with a fuzzy image of a Nazi death's-head eagle embossed upon it. Other pictures showed the 'glass tunnels'.

I started asking my questions. 'You allege', I began, 'that there are signs of this previous civilization, or even extant civilization, on the surface of Mars, which have been detected by certain space probes, and that there has been a conspiracy at a certain level by the powers that be within NASA and other organizations to cover this up.' His reply was, as it turned out, typical Hoagland – an impression of backtracking from the most sensational claims:

> Well, first of all, you've mischaracterized a couple of my positions. I've never said there is an extant civilization on Mars. The imagery we've looked at, the ruins we believe we've detected – we worked through many years with a number of colleagues to decide whether they're real or figments of our imagination – are in fact that: they are ruins, they are very, very old, they look very rugged, they look like they were left a long time ago.

Hoagland's thesis is that NASA, in the early days of its existence, decided that, should evidence of alien artifacts ever be discovered, and publicized, it would lead to grave disquiet among the American public – disquiet that could threaten the very fabric of the nation. So, when evidence *did* start to be discovered, not just on Mars but earlier, on the Moon, there was a powerful effort to cover it up. But what about NASA's announcement that it had found possible microbes in that Martian meteorite? What about the search for water and life? If NASA is trying to hide from the fact there might be life on Mars, surely it is going about it in a very odd way? 'It's

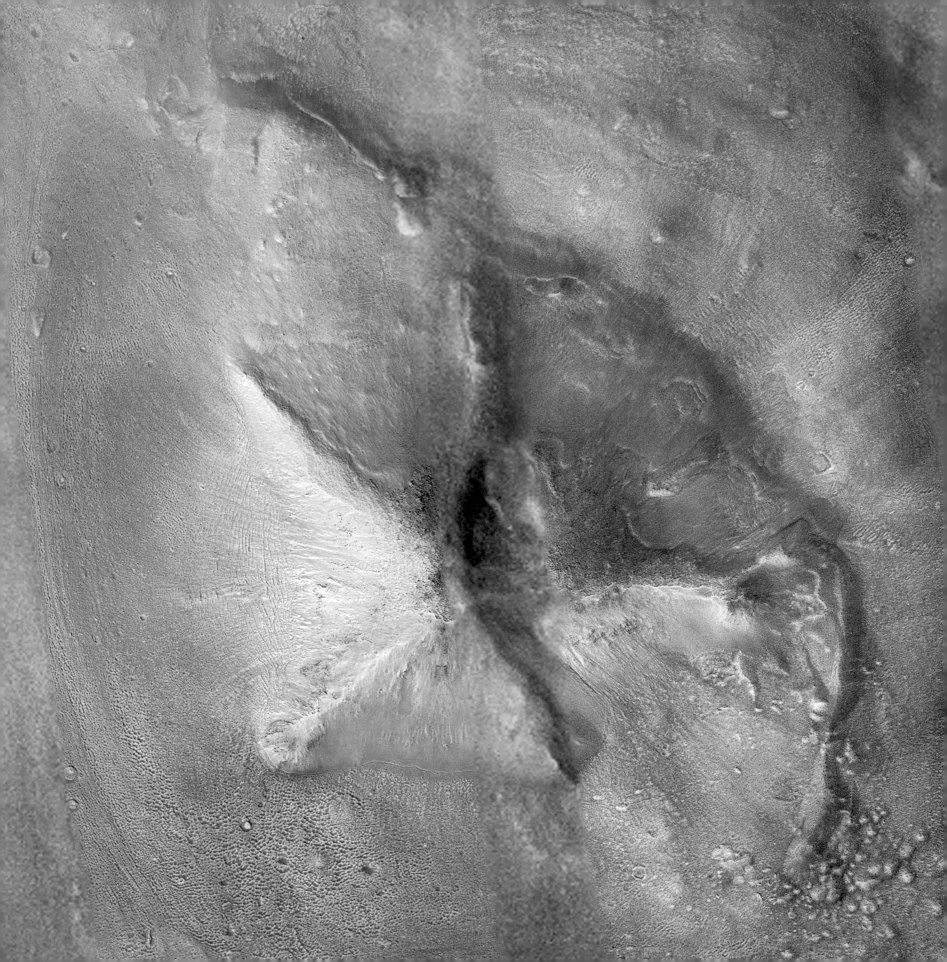

'Microbes are not Martians. Microbes don't hold ray guns, they don't fly spaceships, they don't threaten you. They're little tiny guys, and these, by the way, were long dead'

Richard Hoagland

really very simple,' said Hoagland. 'Microbes are not Martians. Microbes don't hold ray guns, they don't fly spaceships, they don't threaten you. They're little tiny guys, and these, by the way, were long dead.'

When Michael Malin and JPL agreed to use the Mars Global Surveyor spacecraft to retake a picture of Cydonia, there was something of an outcry among the scientific community. Nevertheless, NASA saw it as an opportunity to quash the Face story once and for all. Unfortunately, of course, it didn't quite work out like that. The new, improved 'face' revealed itself to be a scraggy hill surrounded by scree, and the Face conspiracists were not happy. The initial MOC images were poorly defined and showed up badly in terms of contrast. This was hardly surprising given that the probe was not passing directly over the site – to view the 'face' it had to tilt over and view it at an oblique angle. Later photographs showed the 'face' in all its (natural) glory, but the damage had been done: NASA was still trying to cover up what would be the most sensational discovery in history – evidence of an alien artifact on Mars.

Richard Hoagland claims some of the credit for persuading JPL and MSSS to point the camera at the Face, but he claims that in the end it was a reluctant NASA that caved in and went over Mike Malin's head, forcing a man who claims to be so bugged by Face enthusiasts that he is reluctant to hand out his phone number and has some very strong filters on his email. When I asked Hoagland if there wasn't an inherent contradiction in what he was saying – surely, if the existence of the face was something NASA wanted to hide, then why would the Agency then put pressure on one of its contractors to reveal the face? – his answer was simple: infighting within NASA:

I'm saying there are forces, factions. Everyone knows that if your superior tells you not to do something, and you believe in him, you have trust in him, you work with him; you probably won't do that thing. If he makes a reasonable case ... if he says, 'This is a bunch of crackpot nonsense and we don't want the agency associated with, this kind of stuff', you're probably not going to go against his wishes.

Hoagland is clearly passionately committed to his cause, and I came away with the strong impression that, if he didn't believe in all this stuff when he started, he certainly does now. But that doesn't matter; what does matter is that, apparently, quite a few people all over the world believe *him*. Hoagland's ideas fit in with a new, global belief system that encompasses not just lost civilizations on Mars but links to ancient constructions on Earth such as the pyramids. People have always believed weird things, of course, but this fast-moving, Internet-driven conspiracy-mania is something new. It is a fair bet that a fair proportion who take Hoagland's ideas seriously also believe that the Moon landings were faked. Apparently, several tens of millions of Americans take this idea, which is frankly far madder than the idea of an artifact on Mars, seriously. I was prepared to take the whole idea that serious chunks of the population had fallen for the faked Moon landings myth as a myth in itself, but then I kept finding people who swore it was true.

But back to Mars. It turns out in fact that the whole artifacts-spotted-in-space-photos phenomenon is not new, and dates back to the mid-1970s. In 1977, George H. Leonard wrote a book called *Someone Else is on Our Moon*, an illustrated argument that extraterrestrial intelligence of some description was responsible for shaping the lunar surface. The evidence consisted largely of photographs taken from lunar orbit and from the surface by the Apollo astronauts. Fuzzy black-and-white images showing what the author states are 'manufactured vehicles' and 'X-drone rock moving machines' (but which look like, well, rocks) illustrate another roving con-

(*Opposite*) The 'D&M Pyramid'. According to some, in the dimensions of this roughly pentagonal hill lurk numbers of mystical significance. Thanks to the greater resolution afforded by the Mars Orbital Camera, features such as this, which seemed artificial under Viking's low-resolution gaze, resolve themselves into the perfectly natural structures they always were.

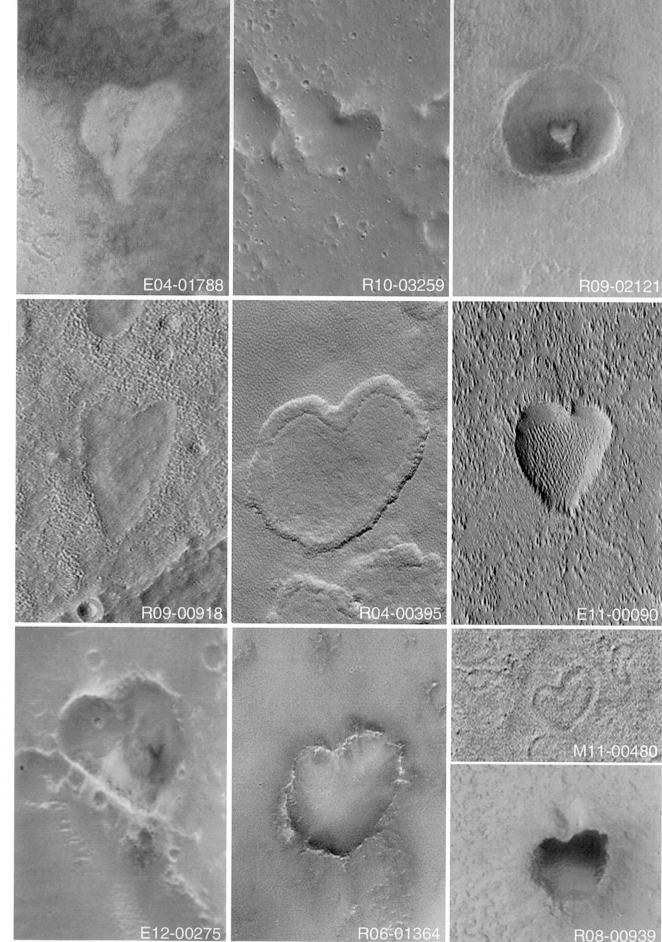

E04-01788

R10-03259

R09-02121

R09-00918

R04-00395

E11-00090

Someone on Mars loves us. This collection of heart-shaped structures on Mars was put together by Mike Malin's team to show that just about anything can be found on Mars if you look hard enough. Of course, there is nothing spooky about these formations, they are just craters, mesas and depressions given meaning by our brains.

E12-00275

R06-01364

M11-00480

R08-00939

spiracy theory, this time involving a secret NASA–Soviet pact to keep the public from finding out that the Moon is inhabited by subterranean super-beings from who-knows-where.

Does anyone in the scientific mainstream believe in faces on Mars? I trawled through the program of the 2003 Sixth International Conference on Mars, looking for any mention of Cydonia. The conference was the height of respectability – no UFO freaks, no dressing up as aliens for the barbecue evening. But then I saw what I was looking for – a poster presentation by one Mark Carlotto. The name rang a bell – Carlotto features heavily in Mars Face discussions on the Web. If Hoagland, with his conspiracies, represents the more colorful end of the Face on Mars community, Carlotto is very much on the gray end of the scale. He has done some mathematical analysis on the Cydonia 'features' and concluded that, in simple terms, there is too much regularity in the landscape to be explained by natural forces such as erosion. I found him standing by his poster, very much alone, and he seemed delighted to have someone to talk to. 'Artificiality is one hypothesis', he said. 'I think we should be open to all possibilities. Mars is full of surprises. I was delighted to get a poster here'.

In a way it is a shame that the Face on Mars debate has all been about aliens and conspiracies, Nazis and Masons (two other of Hoagland's favorite topics). Because, of course, if there were evidence of alien artifacts on Mars it would indeed, as Hoagland claims, be the find of the century, possibly of the new millennium. There is actually no reason why Mars should *not* be littered with artifacts. Most sensible astronomers accept the existence of alien life. It was probably a good thing that Malin was persuaded to take that picture of the Hill that became the Face. Like the recent shambles in which NASA cancelled its commission on a book that was to prove once and for all that Apollo was not a figment of the imagination, ignoring these things and hoping they will go away is counterproductive. If NASA found unequivocal evidence of artificial structures on Mars, and proclaimed it loudly, the public would probably vote to quadruple the Agency's budget overnight. It is hard to see why it would be in anyone's interest to cover something like this up.

In the end, the Face on Mars will just be another item on a long list of myths and legends about this extraordinary planet, a place in which people find so much, frequently on the basis of so little evidence. Clearly, that a great deal of people get a great deal of pleasure from believing in Martian faces, crop circles and lunar cover-ups – life's rich pageant and all that. I am not so convinced they do so much harm. Who knows, maybe some day one of Mars's increasing population of robots *will* find something sensational. It will be interesting to see what Hoagland and his fellow believers say then.

Chapter 9
Global Warming on Mars

'Mars is changing, and it is changing on a time scale that we can measure and observe'

Mike Malin (2001)

Until very recently, it was assumed that whatever Mars was like in the distant past, it has been like it is today for a very long time indeed. But now there seems to be some evidence that not only may Mars have changed its spots quite recently, but it may still be doing so – before our very eyes.

The idea that planets can change dramatically was until quite recently anathema to geologists and climatologists. The driving theory behind geology was the principle of uniformitarianism, in which the surface of our planet owes its existence to a never-ending and extremely slow cycle of uplift and erosion. Geologists believed that no matter how old the rock they were looking at, the features they could see formed in environments not different in essence to the environments we see now. This idea was seen as a logical, rationalist antidote to the religious catastrophism that had been used to explain events and phenomena such as mountaintop marine fossils in the past.

But then along came the new catastrophists. Maybe things had not always been the same on planet Earth after all. The most famous of the great geological catastrophes was the event which took place at the end of the Cretaceous, around 65 million years ago. Along with many other species, the dinosaurs became history. In the 1980s, the belief took hold that the demise of the dinosaurs was triggered by an asteroid, which impacted the coast of what is now Mexico. There were other, less famous catastrophes. Something killed off about 75 percent of all life on Earth at the end of the Permian, 250 million years ago. And 600 million years ago something very terrible seems to have happened. According to some scientists, the whole of the Earth was engulfed in an extreme ice age, what the Norsemen would have called *fimbulwinter*, in which the polar ice caps extended all the way to the equator. The Snowball

Earth was locked in what could have been a permanent deep freeze, but was saved by the volcanic expulsion of greenhouse gases which eventually triggered the mother of all thaws.

With the acceptance of the idea that Earth's past has probably been a great deal more eventful than was previously thought, attention is now being turned to Mars, where things may also be changing quite dramatically. Mike Malin, the man behind the eyepiece of the Mars Orbital Camera, thinks he has found evidence for dramatic Martian climate change – not the ancient climate change that freeze-dried the planet 3 to 4 billion years ago, but climate change going on right now. If he is right, then by the time humans finally set foot on Mars, they may find a rather different planet to the one we see today. 'The present,' Malin says, inverting the motto of the uniformitarianists, 'is not the key to the past.'

His evidence comes from photographs taken near the south polar cap between 1999 and 2001, which seem to show layers of frozen carbon dioxide subliming away. By comparing pictures taken one Martian year (687 Earth days) apart, Malin has discovered that pits and troughs in the dry ice are growing visibly and substantially. 'Owing to the global dust storm that obscured most of Mars during July, August, and September 2001,' Malin says,

(Opposite) Mars in the grip of ice. This image was used on the cover of the December 2003 issue of *Nature* to illustrate an article suggesting that the Martian climate varies quite dramatically over time. In warmer, wetter periods, water would be released into the atmosphere and deposited on the surface again as snow. Paradoxically, a milder Mars would actually experience an ice age, as depicted here.

BUT IS CLIMATE CHANGE REALLY HAPPENING ON MARS?

Global warming is controversial – on Mars as well as Earth. Some scientists maintain that the 1°C (1.8°F) rise in global temperatures on Earth since around 1900 is nothing more than part of a cycle of natural variation. And the popularly portrayed scientific consensus on the likely causes and effects of terrestrial climate change is far from the truth. Martian climate change is less emotionally charged, but scientifically just as hotly debated. Vic Baker, a strong advocate of ancient water on Mars, agrees with Malin on one thing – that the present is not the key to the past. 'In terrestrial geology,' he says, 'the uniformitarian principle has a long tradition of advocacy and also one of scientifically counterproductive outcomes.' But on the issue of climate change on Mars, he and Malin are at loggerheads: 'Malin's theory has been discredited by new data.'

The new data he is talking about point to a Martian south polar cap made mostly of water ice, not frozen carbon dioxide as has been generally assumed. And if the white stuff is mostly ice, that is bad news for the theory that boiling off the cap will thicken the atmosphere. The findings were reported in the 14 February 2003 edition of *Science*. In a paper entitled 'A sublimation model for Martian south polar ice features', Caltech's Shane Byrne and Andrew Ingersoll modeled the growth of Malin's 'Swiss cheese' features – the circular depressions in the south polar CO_2. They found that the key features of the etched pits in the ice – the very flat floors and the steep, rapidly retreating walls, and pits floors which are suspiciously warmer than their surroundings – imply a very thin layer of dry ice, maybe as little as 8 m (25 ft) thick. The floors of the pits are actually composed of water ice, not frozen carbon dioxide, and this is where the etching stops. The Thermal Emission Imaging System (Themis) aboard Mars Odyssey also detected large amounts of water ice around the south polar cap.

Byrne and Ingersoll concluded that there is only a thin skin of dry ice covering a south polar cap that is, in fact, more than 99 percent water ice. At 8 m down – the consistent depth of each of the circular pits – the dry ice stops and the water ice begins. The floors of the pits are too warm to be made of CO_2, they suggest, and so must be frozen water. That explains, says Ingersoll, why the pits never get any deeper:

Water is practically inert on Mars, and so when you sublimate the CO_2 and it reaches the floor and hits the water ice, it can't go any farther for the simple reason that water is much less volatile than carbon dioxide at these very low temperatures.

If Byrne and Ingersoll are right, there is simply not enough carbon dioxide in the polar caps to thicken the atmosphere by a substantial amount – certainly not enough to end the Martian ice age. Even if every gram of dry ice were to be liberated, 'it would change the present atmospheric pressure by no more than a millibar', says Vic Baker. 'Basically, the south polar cap is mostly water ice, so there is not enough CO_2 for climate change.'

VOLCANIC WARMING

Vic Baker and others believe that you do not need to melt the poles to get a warm, wet Mars. Volcanic activity can do the job far more easily. Back in 1991, Baker wrote a paper in *Nature* suggesting that Mars undergoes episodic spasms of volcanic activity, which would have dramatic effects on the climate of the planet. 'I believe that volcanic and hydrological processes, acting together, may result in climate change', he says. 'The release of planetary heat melts ground ice and releases volatiles in addition to water – gases like carbon dioxide. My model produces episodic climate change, occurring within the last ten million years.' Baker sees evidence that water flows on the Martian surface today, not in Malin's gullies, but in one of the most puzzling features ever seen on Mars's surface – thousands of bizarre dark streaks that were first glimpsed by the Vikings in the mid-1970s.

At a presentation at the Lunar and Planetary Science Conference in Houston in March 2003, Baker and others argued that many of these mysterious features could be caused by water flowing across the Martian surface. Justin Ferris, a colleague of Baker's, wrote a paper in 2002 suggesting that the dark streaks could have been caused by water, but publication (in *Geophysical*

(*Opposite*) All over Mars are thousands of enigmatic dark streaks, which can be seen tumbling down the walls of craters and other cliffs. Some geologists have suggested that we are seeing active liquid water flows in action; others say they are just dry avalanches of dust.

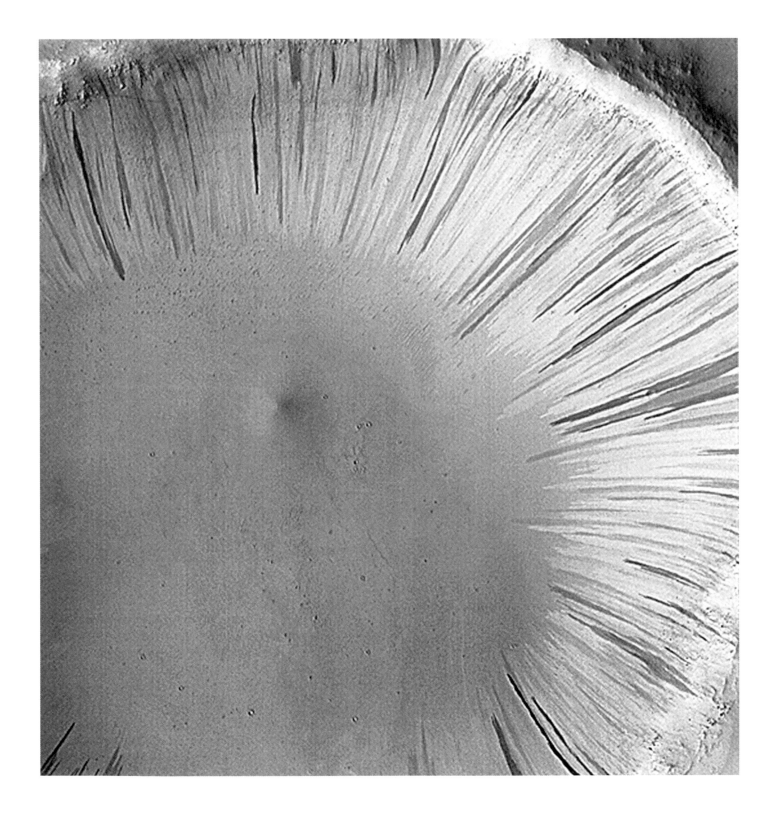

Some of the dark streaks seem to fan out over a very wide area, suggesting that whatever they are made of is very fluid – possible extremely fine, dark-coloured dust. The suggestion that water – or some other liquid – is involved in the formation of some or all of the streaks, is highly controversial. This picture was taken in June 2003 by the MOC and shows a crater, just north of the Martian equator, whose walls have been marked by avalanches of dark material approximately half a kilometer across.

Research Letters) was held up for nearly a year because reviewers were adamant that water could play no role. 'Now that numerous other features are being tied to recent near-surface-water-related activity,' Baker says wryly, 'those reviews can be seen as an impediment to scientific progress.'

There are, it seems, two types of dark streak. Some are narrow and straight, and according to Baker and Ferris their formation may involve nothing more exotic than mass wastage – basically rocks and soil rolling down a slope. 'But what intrigued us', writes Baker, 'is that some of the streaks are *anabranching* – braided – spread out over a wide surface, their shape depending on angle of slope. Water is consistent with some of the forms we are seeing.' The streaks are probably the youngest features we have yet seen on the Martian surface. None are cratered, and all overlay other surface features. They appear to be especially common near the Tharsis volcanoes, suggesting that geothermal warming may be responsible for liberating groundwater from permafrost. Tahirih Motazedian, a geologist at Oregon University, thinks that in these streaks we have the proof that Mars is still geologically active. 'It is possible that geothermal activity driven by volcanic heat melts subsurface ice, releasing a brine that dissolves subsurface materials.' In a paper presented at the 2003 Lunar and Planetary Science Conference, she wrote:

> This brine has a low freezing temperature, allowing it to flow at the Martian surface. The dissolved minerals precipitate from solution, leaving behind dark streaks of rock varnish. The dark streaks appear where the water table is intersected, as in craters and valleys.

In a paper published in *Geophysical Research Letters* (12 December 2002), Oded Aharonson and colleagues made a strong connection between the streaks and water. After studying 23,000 images, they found that streaks occur only on steep slopes and, most importantly, in places where peak temperatures regularly exceed 2°C (36°F). The streaks that form in the coldest environments tend to occur on south-facing slopes in the north. At the March 2004 Lunar and Planetary Science meeting, Allan Treiman and Michel Louge suggested that the streaks have a more mundane origin – they are simply dry avalanches of 'thin layers of bright, wind-deposited dust exposing darker sand or rock beneath'.

One explanation offered for the Martian gullies is that they form under snowpacks (arrowed), in effect mini-glaciers. Liquid water trickles out from their undersides, eroding the loose regolith and rock beneath.

THE GULLIES: IMPLICATIONS FOR MARTIAN CLIMATE CHANGE

Until recently, there were three competing theories for the formation of the Martian gullies. The first is that they are formed by liquid water or brine gushing out from the crater walls. The second is the exotic idea that the gullies are formed by liquid carbon dioxide. The third possibility is that loose material is carried downslope suspended on a bed of carbon dioxide – part of Nick Hoffman's White Mars ideas. But there are problems with each of these. Liquid CO_2 requires pressures of several bars to be stable, so if it were to exist in Martian rock pore spaces some mechanism would have to be found to keep these potentially explosive strata isolated from the atmosphere. The gullies – as opposed to the dark streaks – are mostly found at high latitudes and away from any plausible sources of volcanic heat. It is hard to see how they could have been formed by springwater – the rocks are simply too cold.

In February 2003, a paper published in *Nature* by Philip Christensen entitled 'New model for gullies on Mars' contained a wholly new theory of how the gullies formed. Christensen thinks that the gullies were created by trickling water from melting snowpacks, not underground springs or pressurized flows of either water or carbon dioxide. The snow provides a source of water, and also protection for the unstable liquid in the thin atmosphere. In a THEMIS image, Christensen saw eroded gullies on a crater's cold northern wall and immediately next to them a section of what he calls 'pasted-on terrain'. This terrain looks like, well, dirty snow that has filled in the grooves in the crater wall. Scientists have known about this terrain for some time, and have assumed that it is composed of some sort of volatile substance because it is found only in the coldest, most sheltered areas. The most likely composition of this material is snow. Christensen suspected a special relationship between the gullies and the snow.

'I saw it and said, "Ah-ha!"', recalls Christensen. 'It looks for all the world like these gullies are being exposed as this [pasted-on] terrain is being removed through melting and evaporation.' He continues:

The gullies are very young, that's always bothered me. How could Mars have groundwater close enough to the surface to form these gullies, and yet the water has stuck around for billions of years? Second, you have craters with rims that are raised, and the gullies go almost to the crest of the rim. If it's a leaking subsurface aquifer, there's not much subsurface up there. And, finally, why do they occur preferentially on the cold face of slopes at mid-latitudes?

That is the coldest and the least likely place for melting groundwater. Christensen says that trickling snowmelt solves most of these problems:

Snow on Mars is most likely to accumulate on slopes that face the north or south poles – that is, the coldest areas. It accumulates and drapes the landscape in these areas during one climate period, and then it melts during a warmer one. Melting begins first in the most exposed area right at the crest of the ridge. This explains why gullies start so high up.

THE MARTIAN ICE AGE: A TOPSY-TURVY WORLD

Here on Earth we live in an age of ice. It is easy to forget, in the current furor about climate change, that right now the Earth is colder and icier than it has been for most of its history. For much of the Mesozoic era, for example, when the dinosaurs trod the Earth, the average temperature was significantly higher and there was little permanent ice on the planet at all. Antarctica, even when it had drifted to roughly its current polar location, remained ice-free all the year round, and indeed was covered with forests (hence the coal deposits that we find there today). Ice ages have occurred sporadically throughout Earth's history – right back into the Precambrian. A series of super ice ages – the so-called Snowball Earths – may have dominated the early Cambrian, and could even have been linked to the explosion in evolutionary diversity seen at that time in the fossil record.

But when we think of ice ages, we think of the recent glaciations, the last of which ended just before recorded history, around 8,000–10,000 years ago. For the last couple of million years, Earth has seen a series of retreats and advances of polar and mountain ice triggered and accom-

This Themis image shows dozens of 'snowpacks' packed around a crater wall – areas of light-coloured material, probably 'dirty' ice, packed into the dozens of gullies and alcoves.

panied by global temperature changes of up to 10°C (18°F). The causes of these ice ages remain unclear, but they appear to be linked to cyclical variations in solar output, in the eccentricity of Earth's orbit, and in the angle of its axial tilt. If Earthly ice ages can be triggered by cosmic events, then presumably the Martian climate can be altered by similar events.

In December 2003, an intriguing article entitled 'Recent ice ages on Mars' appeared in *Nature* suggesting that Mars may experience ice ages. Jim Head, of Brown University, and colleagues suggested that Mars too has experienced recent glacial and interglacial periods corresponding to terrestrial glaciations. Head has identified what he believes are glacial landforms, snapped from orbit by the Mars Orbital Camera, landforms that were created very recently – perhaps as recently as 400,000 years ago. These features include layered sedimentary deposits a few meters thick, small-scale 'polygons' similar in size and shape to the ice-wedge phenomena of Earth's polar regions, the famous gullies, and, most convincingly of all, what look like the remains of old glaciers, now buried under dust.

Head believes that the most recent Martian ice age may have finished very recently, perhaps as recently as half a million years ago or less. During Martian ice ages, which occur during periods of high axial tilt (greater than 30° or so), ice is removed from the pole regions and deposited as frost and snow at lower latitudes. 'If such an ice cover had occurred on Earth, it would have reached southward to latitudes equivalent to Saudi Arabia, North Africa and the southern United States', wrote Head.

But Martian ice ages are nothing like those on Earth. In fact they seem to be totally the opposite in cause and effect from what our planet experiences. For a start, Earthly ice ages are, by definition, cold. Martian ice ages require the poles to warm up to release their store of frozen water into the atmosphere. Vic Baker, commenting on the research, wrote in *Nature*:

For Mars, the episodes of ice accumulation ('glaciations') are warm phases for a planet that seems to be most stable in an ice-house condition. For Earth, the stable sate seems to be the warm greenhouse condition, with cold phases corresponding to metastable periods of glaciation.

According to Jim Head, 'glacial periods on Earth are characterized by colder temperatures at the poles on average, while on Mars the reverse is true.'

So, Martian ice ages might be exactly the reverse of Earthly glaciations – they are warmer and 'wetter' than the cold, dry periods such as we see now. It is not at all clear what could be causing this climate change, or whether what triggers an ice age on Earth might be related to or be the same as what causes a cold dry period or ice age on Mars. Baker believes that the causes of climate change on Mars may well be related to volcanic activity – it is bursts of intense volcanism that he believes were responsible for the periods of warm, wet fluvial activity in Martian deep time. Cosmic effects may play a role too, particularly changes in the density of interplanetary dust, in turn related to the solar system's periodic passages through the dusty spiral arms of our galaxy (a postulated mechanism for inducing terrestrial ice ages). As he says, 'With two planets on which to explore this puzzle, we should have a better chance of resolving it than we would have by limiting our investigations to only one.'

What a long way we have come from the cold, dead ancient world of Mariner 4. Climate change, a dirty phrase on Earth, symbolizes the new Mars, a dynamic world where things can get better, in the sense of becoming more like the Earth. If Mars is undergoing some sort of climate change related to cosmic effects, then finding out more about it should be an essential scientific goal. After all, if changes in the Sun, say, were found to be driving climate change on Mars, the same changes could probably be linked to global warming on Earth – a heresy today, but a reasonable hypothesis that too many are too quick to damn. But however exciting Mars gets for the scientists, there are some people for whom the latest data from Mars Global Surveyor, Odyssey, and the Mars Exploration Rovers will just never be enough.

Chapter 10
The Dog that Didn't Bark

'It's like sending somebody a love letter. You know they've got it, and you are just waiting for the reply'

Colin Pillinger (Boxing Day, 2003)

Never has the agony of robotic space exploration been as poignantly felt, or been on such public display, as during the Christmas and New Year holidays at the end of 2003, when the Beagle 2 probe to Mars became one of Britain's 'Great Heroic Failures'. Beagle was clever and had a great name. The Beagle team trumpeted the fact that, unlike their rivals on what the press liked to imagine was the 2003 race to Mars, the wheeled Mars Exploration Rovers, the British probe was equipped with a spectrometer capable of detecting compounds – namely methane and certain isotopes of carbon – that would signal either extant or past life. Unlike the JPL machines, Beagle was small and contained engineering of unprecedented minuteness and precision, forced upon the probe by the extremely low weight limit of 69 kg (152 lb) imposed by the European Space Agency (ESA), whose Mars Express mission got it to Mars. Beagle must be the only interplanetary spacecraft that could be lifted by a man.) And also unlike the JPL machines, Beagle was almost absurdly cheap – a £35 million bolt-on to the £50 million Mars Express. Beagle was also British, from the same land that gave the world, the steam engine, the Spitfire, the practical computer and Concorde. In the shape of its creator, Professor Colin Pillinger, Beagle 2 had a media-friendly front man, the archetypal eccentric British scientist. With his mutton-chop whiskers, a day job as a farmer, and a rural accent, Pillinger is as far removed from the image of the JPL rocket scientist as you could get.

The Beagle project began in the quietest of ways in 1996, with (Pillinger insists) a doodle on the back of a beer mat – showing a Mars lander design that could hitch a ride on the Mars Express orbiter due to be launched in less than six years' time. Pillinger, a scientist at the Open University, was a world-leading expert in the study of the small group of meteorites known to have come from Mars, and he cut his research teeth in the 1970s on a series of investigations of lunar rocks. The concept of an instrument to search for life itself – as opposed to the circumstantial evidence for life – originated with the European Exobiology Study Group in the mid-1990s. When the Mars Express mission was given the final go-ahead, Pillinger pushed for a lander to be included as part of the package.

Although a British project, Beagle was more or less ignored in its home nation until the last moment, when it generated a flurry of interest in the British media not shown in a space mission since the days of Apollo. Unlike NASA missions, in which every last cent must be accounted for in forensic detail, Beagle's finances consisted of wings and prayers concealed behind smoke and mirrors. No one ever owned up to how much it cost – £35 million was the on-message estimate, but even the gentlest of probings shows that the true cost was £50 million or more. Pillinger knew where the money came from, but all the sources were never revealed. Some was squeezed out of the British government, some from ESA, and some came in the form of work performed gratis by the main contractor, Astrium. Much must have been borrowed, and it is to be hoped that these debts will now be written off. Early on, Pillinger talked of winning sponsorship from big names. British Telecom and Virgin, to name two, were contacted, but both

(*Opposite*) The arrival of Beagle 2 on Christmas Day 2003 caught the British public's imagination as had no other space mission since Apollo. This was the **London** *Evening Standard*'s front page, a couple of days before touchdown, showing an artist's impression of Beagle's fiery plunge through the Martian atmosphere. The failure of Beagle was greeted with a mixture of intense disappointment and the sort of self-flagellation – 'another British failure' – that the UK does so well.

THE BEAGLE IS LANDING

Top scientist says British Mars probe has excellent chance of finding life

BRITAIN'S Beagle 2 probe to Mars was hours away from landing today ... with the man behind the mission confident it will succeed in finding signs of life on the Red Planet.

Professor Colin Pillinger told the Evening Standard: "This is the first mission to look for life on Mars, rather than just signs that life may have existed in the past. We would not be doing this if we didn't think there was a good chance of success."

Full story: Pages 4 & 5

Nearing touchdown: the British Beagle 2 is shown in this artist's impression approaching the surface of Mars to begin its search

167

declined to be associated with a mission to a planet which had claimed the lives of more than half the robots which had been sent there. Beagle was clever and audacious, and if it had worked would have shown the value of exploring space quickly and cheaply – the ultimate expression of Dan Goldin's philosophy of faster, better, cheaper. But sadly, of course, while Beagle 2 may have been faster and cheaper than any other space probe in history, it didn't work. We shall probably never know the real reasons, unless the beady eye of NASA's Reconnaissance Orbiter spots it lying on the frigid plains of Isidis, where it is presumed to have landed. This would at least provide some clues. The sorry tale of Beagle is one of what might have been, and a salutary reminder that going into space is both extremely hard and extremely risky.

Beagle 2 was not the first planned mission to the Red Planet to use this name. In the early 1990s, Bob Zubrin, of Mars Direct fame, had also christened his putative manned voyager to Mars after the vessel in which Charles Darwin undertook his voyage of discovery in 1837. On his world voyage, the English naturalist had observed similarities and differences between species, and years later these observations crystallized in a theory that revolutionized not only biology but the very way we look at ourselves. Pillinger's Beagle 2, it was hoped, would do something of the same. Forged very much in the spirit which believed that life on Mars was at least a strong possibility, Beagle was designed with one aim: to find life. To this end it came equipped with a set of instruments that would probe, scrape, scratch, burrow into, and cook bits of Mars, hoping to smelt out the telltale signs of a second genesis. Beagle would not have been the first probe to search for life on Mars. Both the Viking landers of 1976 were equipped with mini-laboratories to test for the products of organic chemistry, but the results were considered negative by some, inconclusive or ambiguous by others. Beagle's results, when they came, were supposed to be definitive.

Beagle's star turns were its Position Adjustable Workbench, a pretty meaningless label chosen purely for its acronym, and the Mole, a clever burrowing device that would hammer itself

Colin Pillinger, the public face of the Beagle project. With his Gloucestershire accent and extravagant facial hair, he was the epitome of the Great British Eccentric, and epitomized a mission that got off the ground on a wing and a prayer.

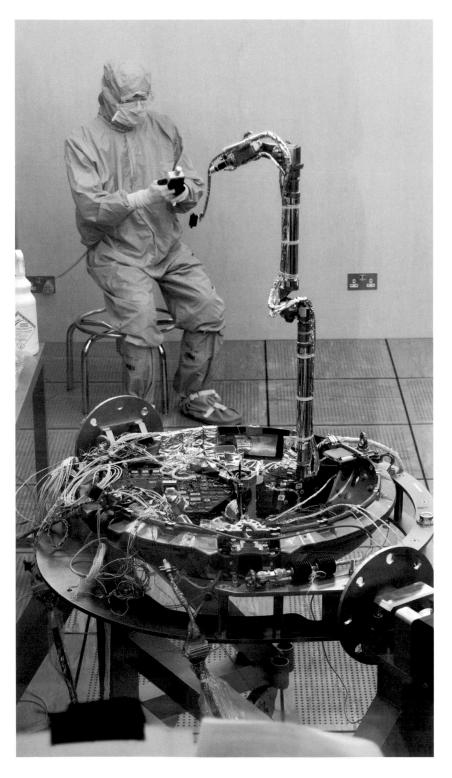

down into the Martian regolith to a depth of a couple of meters or so, extract a soil sample, and return it to the lander, where it would be cooked and analyzed by the miniature gas spectrometer. The Gas Analysis Package, which combined the functions of no less than four mass spectrometers and gas processing systems, was a masterpiece of baroque miniaturized engineering, matching the genius of the best Swiss horologists. For example, twelve tiny platinum ovens were constructed, housed on a miniature carousel, each designed to receive a sample of Martian soil or rock and sealable with a sapphire bung. Normally, equipment like this would be the size of a small room, or at least a cupboard; the Beagle team had to make it fit inside a volume about the size of this book. Beagle was equipped with stereoscopic vision to peer at its surroundings, and with the equivalent of a geologist's microscope to examine Martian rocks in close-up. There were also rock grinders and machines for sniffing the composition and pressure of the atmosphere. As an exercise in taking an entire laboratory and shrinking it by perhaps two orders of magnitude, Beagle 2 was an outstanding success.

The key experiment was to try to detect 'isotopic fractionation' between different carbonate minerals. Organisms sift out and concentrate certain isotopes of common elements, carbon in particular, so that organic matter, whether living or dead, tends to contain slightly different ratios of these isotopes than does purely inorganic material. (Different isotopes contain different numbers of neutrons, but are identical in terms of protons and electrons. The resulting difference in mass gives different isotopes of the same element a slightly different chemical behavior.) For example, living things tend to concentrate the proportion of carbon-12 over its other biologically sensitive isotope, carbon-13. Thus you end up with a ^{12}C:^{13}C ratio ('delta ^{13}C' in exobio-speak) that is different to that produced by inorganic processes.

Beagle's science package was clearly second to none. The problem lay, as with all such missions, in getting this package to Mars and safely down to its chosen landing site – Isidis Planitia,

Many people gave their time for free to make the Beagle 2 project a reality. Not that there was anything amateurish about the design or the methods used for its construction. It is likely that elements of the Beagle science package will be used on future missions, even if the spacecraft itself is never rebuilt.

just north of the Martian equator, a flat and relatively rock-free area marked by a series of small volcanic hummocks and with a high subsurface concentration of ice, as indicated by the neutron spectrometer aboard Mars Odyssey. The first part of this objective was accomplished without fuss. Beagle sailed aloft, piggybacked on board the 12-metre-wide Mars Express Orbiter, Europe's collaborative and rapidly deployed replacement for Russia's failed Mars 96 mission (minus one of the landers and the penetrator-probes) which ended shortly after take-off in that year.

From Baikonur, Mars Express plus Beagle blasted into the Kazakh skies on 2 June 2003 aloft a Soyuz-Fregat launcher, a cheap, reliable but rather small rocket. The 400-million-km (250-million-mile) voyage to Mars was uneventful, apart from an early power-glitch which seems to have had no long-term consequences for the mother ship, and a solar storm which buffeted the inner solar system shortly before Beagle's spin-up and eject mechanism detached the lander from its mounting on 19 December 2003. This event was marked by a party in London at which there was both a ministerial and royal presence. Poignantly, the last-ever picture of Beagle was obtained a few minutes after separation. It shows the little probe, shaped like a cartoon flying saucer and about as big as a dustbin lid, drifting away to its fatal encounter with Mars.

If the science package aboard Beagle was a triumph of miniaturization and ingenuity over the constraints of space and mass, the lander itself, together with its various descent systems, was a triumph of expediency over the desirable. From the point at which Beagle left the orbiter, it was on its own. In the event, the trajectory appeared to have been perfectly calculated, and there is no reason to suspect that Beagle struck the Martian atmosphere at anything other than the optimum velocity. Beagle's descent was to be slowed by three means. First, a cork-faced heat shield measuring about a meter across would allow Beagle to aerobrake down from a speed of around 20,000 kph

Beagle 2 successfully separated from Mars Express on 19 December 2003, and a party to mark the occasion was attended by British royalty and members of the government.

(12,500 mph) to around Mach 4 (1,200 kph/750 mph on Mars), when a pilot parachute would deploy, fired out of the plummeting spacecraft by an explosive cartridge. The opening of the pilot chute would also release the heat shield from the rest of the lander. The pilot chute would slow the probe down to around 335 kph (210 mph), when the main parachute would open – a 10 m (33 ft) ring-sail canopy, described by just about everyone who knows about these things as the finest parachute ever constructed. This was designed to slow Beagle down to 58 kph (36 mph), when its radar altimeter trigger (RAT) would have triggered the deployment of Beagle's triple-airbag system at about 200 m (650 ft) above the ground. On the signal from the RAT, the three orange-segment-shaped, ammonia-filled bags were to inflate to cushion the lander as it struck the surface.

It was calculated that Beagle (provided the landing site had been chosen carefully, for flatness and absence of too many sharp rocks) should come to rest after about fifteen long bounces. The gas bags would then spring away from the lander, allowing the clamshell structure to open and the solar cells to deploy. Now Beagle could get to work, after sending back to Earth a small image of its surroundings, together with a 'call sign' – a nine-note ditty composed by the British rock band Blur. These would be sent via the radio transmitter aboard NASA's Mars Odyssey

(*Opposite*) A working model of the complete Beagle 2 spacecraft was tested at Sandy Quarry in Bedfordshire. The ingenious Mole, visible at the left, was a tethered burrowing device designed to dig its way down more than a meter to retrieve a sample of soil, which would then be tested for signs of life. The Position Adjustable Workbench (PAW) can be seen bringing its array of instruments to bear on a nearby rock.

'Please don't go away from here believing we've lost the spacecraft'

Collin Pillinger

spacecraft, kindly made available to the Beagle team by JPL as a relay transmitter until Mars Express could be brought on-line in early January. All this was supposed to happen in the early hours of Christmas Day, and the whole event witnessed live at what became known as the media's mission control, at the Open University building in Camden Town, north London.

The Beagle landing party started out as a happy, red-eyed occasion. The auditorium and meeting rooms were packed with journalists, TV crews, and scientists giving catatonic interviews throughout the night. The occasion was even graced with celebrity. Someone from Blur was there, and a member of a girl band no one could place. It was all terribly British: tea and biscuits, nowhere to sit, bright lights, and the uncomfortable furniture of academia.

Then, shortly before 6 a.m. on Christmas morning, we were all hustled into the auditorium. There, a large computer display was set up to show the first pictures from the Martian surface. Pillinger was on the phone, awaiting news of his baby. He looked like a schoolboy about to open his exam results. On the other end of the line was someone at JPL listening in for the signal from Odyssey. At about 6.03, Pillinger announced, that 'Odyssey is now communicating with Earth.' Then, with the phone clamped to his ear, he fell silent. Five long minutes later he put the phone down. 'Unfortunately we do not have any radio telemetry on this pass', he told us. 'It is a bit disappointing but it is not the end of the world. Please don't go away from here believing we've lost the spacecraft.' He was, as it turned out, wrong.

At that point, Pillinger looked flat and rather drained. There was as yet no agony of certain failure, but the possibility had clearly crossed his mind. But all those there were assured time and time again that all was not lost. Getting a signal from Beagle at this first opportunity was never going to be more than a bonus. There would be more passes by Odyssey – once or twice

(*Opposite*) No one doubts that Beagle 2 was a triumph of design and miniaturization. Had it worked, it could have accomplished what many NASA missions have done, but at a tenth of the price. Sadly, it seems, Beagle 2 took cost-cutting a step too far.

a day – and the large radio telescope at Jodrell Bank would, from that night, be using its 76 m (250 ft) dish to scan the whole face of Mars for a signal from Beagle. In January, the Mars Express mother ship would be maneuvered into an orbit from where it would be able to communicate directly with Beagle. The two spacecraft shared a dedicated communications link. Perhaps there were glitches, mismatches between the American and British software aboard Beagle and Odyssey, and they were simply not speaking the same language. Pillinger was upbeat about the failure of Beagle to communicate via Odyssey, but behind the scenes his team was far more depressed. 'I don't care what kind of spin they want to put on it,' one engineer told me that Christmas morning, 'but the fact is that we had maybe an 80 percent chance with that pass. Frankly, it's not looking good.'

He was right. Over the coming days the press conferences became sadder and more thinly attended. At each briefing a list of possible reasons for the lack of signal was discussed. An early theory was that Beagle's clock had been reset by the shock of landing. This could mean that its transmitter was broadcasting at times when Odyssey was not overhead to receive its signal. That possibility was later discounted. Another theory was that Beagle was lying on its side, or at least at an awkward angle, with its transmitter pointing sideways at the horizon rather than straight up at the sky. This too was discounted after several passes by Odyssey at a variety of angles. Then someone threw in the idea that Beagle had fallen into a crater, which had been (somewhat mysteriously) 'spotted' in MGS high-resolution images of the landing site. It was a good line – Beagle swallowed in a hole – and it maintained the media interest, but as an explanation for what had happened it didn't add up. Martian craters do not have such steep-sided walls, and there is little chance that being at the bottom of one would have impeded radio communications. It is possible that Beagle could have been damaged by tumbling down a crater wall, but more likely was that it had hit some of the spiky ejecta found around the rims of such structures. In fact, the crater turned out to be a red herring.

Watching Colin Pillinger and his colleagues try to explain the lack of news from Mars became a painful experience. Clearly, spirits were sinking by the day. The champagne bought to celebrate the triumphant first signal was locked in its cupboard and quietly forgotten. By the time of the landing of Spirit, the first of the NASA rovers, the atmosphere in the Beagle press room was grim. Only one pop star was left, and the BBC man was muttering about taking his broadcast van away. The first pictures from Spirit were welcomed by the Beagle team, but in their heart of hearts they must have seen, as each image rolled in from the Gusev crater, the coffin-lid slowly closing on their baby. For those pictures meant not only that Spirit was in fine health, but the radio aboard Odyssey had not malfunctioned. The failure to get a signal from Beagle from Jodrell Bank was probably the surest indicator that something had gone fatally wrong. Once Beagle had slipped into Communications Search Mode, as it was programmed to do if no contact had been made for a specified amount of time, Jodrell Bank should have easily been able to see it.

A problem with Odyssey's transmitter remained one of Beagle's best hopes. Even then, Pillinger was insisting that he had not given up on Beagle 'in any shape or form'. And he probably meant it, even if his body language was starting to suggest the opposite. It was all very different from the scenes that greeted the failure of the Mars Polar Lander in 1999. Then, another Christmas was ruined, this time for a team from JPL. Then there were real tears when the signal failed to materialize, but at least for the Americans the agony was over fairly quickly. Polar Lander had its own high-gain transmitter, and its failure to communicate with Earth meant that something catastrophic must have happened. After three days it was game over (although a small group of engineers continued to play around with the options for three months or so after the mission was officially declared dead).

But for Pillinger, mission manager Mark Sims, and the rest of the 500 or so Beaglers, the agony went on. The coup de grace came on 7 January 2004, when pressmen, TV crews, and presenters again crammed into the Open University building for what was to be Beagle's swansong. A live Internet linkup with Mars Express mission control in Darmstadt had been set up. Pillinger had again bravely put himself up as the front man for what he must have known by then was a doomed enterprise. With a hundred cameras trained on his face, he listened as David Southwood, ESA's chief scientist, told us what had happened when Beagle 2's mother ship had made its first pass overhead. It must have been an awful moment for Pillinger, watching the screen, like the rest of us, and listening to the news. He didn't even have the luxury of hearing the results in private first, then composing his thoughts and coming out to make a statement. Dr Southwood's first word was 'unfortunately', and the whole room groaned. At that point it was clear that Beagle was lost. No

signal had been picked up from the lander, no science, no pictures. In all probability, those delicate, ingenious instruments – the robot mole, the tiny spectrometers, the microscope, the camera – had been smashed to pieces and strewn across the unforgiving sands of Isidis. By February 2004, all hope for Beagle had been abandoned.

WHAT WENT WRONG?

Beagle *could* have worked. If it had, it would have been feted as a great British triumph. A successful Beagle would have been a marvelous coup; even a picture from the surface and a bit of gas analysis would have counted as a staggering success. It would have proved that you could go to another planet and conduct some serious scientific research – who knows, maybe even discover alien life for the first time – for the price of a plutocrat's mansion in London or New York. And it *should* have worked. Every system on board was tested time and time again. But there were inbuilt and structural flaws with the whole concept that in hindsight (a great perspective on any failure) prejudiced the mission from its inception.

For a start, Beagle had zero redundancy in its descent and landing system. If any one of its crucial landing systems failed, Beagle was doomed, no matter what else was done. The Mars Exploration Rovers (MERs) each had twenty-four inflated airbags to cushion their fall. If one or two of them burst while being inflated, it would not necessarily have ended the mission. Beagle, though, had just three. A single airbag failure would have spelt disaster. Beagle had no retro-rockets, as were used on the MER and Viking landers to further slow or correct the course of their descent. On the MERs, the rockets and airbags complemented each other. Beagle had no exterior transmitter capable of sending a signal to Earth during its descent. Such a device would not have saved Beagle, but it could have provided valuable and

(*Opposite*) A graphic showing the Beagle 2 descent sequence. The enquiry into Beagle's fate released its findings in May 2004, when it was judged likely that a failure in the descent and landing systems was the most likely cause of the loss. The Martian atmosphere was probably significantly thinner than was thought, meaning that Beagle's descent was faster than planned – leading possibly to impact uncushioned by airbags.

Landing	Mission phases	Activity	Distance from landing site	Velocity
-5.75 days			1.5 M km (0.84 M miles)	
-2.5 hours		Spin up and eject		5.4 km/sec (4.5 miles/sec)
		Wake up computer		
-7.7 minutes	ENTRY	Deceleration rate begins to rise	120 km (75 miles)	20,000 kph (12,500 mph)
		Peak temperature 1700°C	34.6 km (18.7 miles)	
		Deceleration rate begins to fall (13.3 g)	24.7 km (15.3 miles)	12240 kph (7650 mph)
		Deceleration (0.8 g)		
-4.5 minutes	DESCENT	Deploy pilot parachute	7.1 km (4.5 miles)	1174 kph (720 mph)
-3.75 minutes		Deploy main parachute	2.6 km (1.6 miles)	335 kph (210 mph)
		Switch on RAT		
		RAT detecting surface every 0.1 seconds		
-15 seconds		Inflate gas bags gas bags filled	275 m (850 feet) 200 m (750 feet)	56 kph (36 mph) vertical 129 kph (8 mph) lateral
	LANDING	First bounce (15 seconds) Sixth bounce (10 seconds) Tenth bounce (6 seconds)		
140 seconds after inflation		At rest (Isidis Planitia). gas bag separation		
		Transfer to lander software		

timely information about what went wrong. Finally, there was only one Beagle.

In fact, there were no fewer than nine 'single-point failures' built into the Beagle separation and descent sequence alone. A string of operations, from the successful ejection of Beagle from Mars Express, to the correct opening of the parachute, the ejection of the heat shield, the airbags, and so on, all needed to work perfectly, first time, for Beagle to get to the ground in one piece. I spoke to one of the men from Astrium, who told me that each of these systems was rated to have a 95 percent chance of working properly. Fire the airbags twenty times and only once will they burst, open the parachute twenty times and only once will it rip or wrap itself around the plummeting probe. Ninety-five percent sounds pretty good, but when you have nine such single-point failures in a row, the odds of success are cut to 63 percent. And that is just for getting down to the surface of Mars in one piece. Add to that the possibilities of a launch failure, of a problem with navigation on the way to Mars, of some glitch in the solar cells, or any of the dozen other hazards that Beagle faced from on the moment it left the launchpad, and the proba-bility of a successful mission probably drops to a mere 50–50.

Colin Pillinger is an adherent of the 'faster, better, cheaper' philosophy first espoused by NASA administrator Dan Goldin in the mid-1990s; the idea of building small, relatively unambitious spacecraft quickly and cheaply using a mixture of bespoke and spare-part components wherever possible. The downside of faster, better, cheaper, say the skeptics, is that your spacecraft is not as comprehensively engineered as a 'Rolls-Royce' space vehicle such as Viking or Voyager. But the upside is that you can perhaps build several such craft and fire them at your target, hoping that this scattergun approach will tilt the odds in your favor. With Beagle, although it was built fast and cheaply, there was not this ultimate redundancy – a second machine as a backup. In retrospect, it would have been better to build two or even three Beagles and fly them to Mars bolted to another mission – after all, the whole point of Beagle was that it was small and cheap. There were also severe tensions between the team responsible for Mars Express and the team that built Beagle. To the Mars Express people, Beagle was never more than an instrument. To the Beagle team (and here they were correct) it was a spacecraft in its own right. Mars Express was never designed with a lander in mind – Pillinger had to bully his way on board. Beagle was built extremely quickly, even more so

than the mother ship, and so much so that certain key systems, notably the radio link between Beagle and Mars Odyssey, could not be fully tested.

What actually killed Beagle is not known. One or more of Beagle's airbags could have failed. If so, the lander would have smashed itself to bits on Isidis Planitia. Almost everyone I spoke to who had anything to do with the project agreed that the airbags were the weakest link in Beagle's landing system. For a start, airbag landing systems are not tried-and-tested technology. They had worked once, on Pathfinder. During tests for Beagle, the airbags leaked and burst and had to be redesigned, and in the rush to complete the spacecraft and have it ready for delivery to the main orbiter the final airbag design could not be fully tested. It is also possible that the parachute was incapable of stopping Beagle from swinging excessively during its descent. Maybe the parachute failed to deploy, or got itself wrapped round the probe. Or the fragile material could have been weakened by the radiation blasting it got before launch as a sterilization measure. Possibly, and simply, maybe everything worked just fine and Beagle simply hit a large, jagged rock – the sort of accident that haunts every robot planetary lander mission and which may in itself argue for always flying a backup. But one intriguing possibility was suggested by JPL's Mark Adler, the day before the landing of Spirit. He announced that Spirit's parachute was retimed to open a trifle earlier than planed, as a recent dust storm on Mars had warmed the atmosphere at high altitude, thinning the air appreciably. Such a warming would have lowered air resistance, and Beagle would have been falling more rapidly than its designers calculated. The opening of the parachute and the firing of the laser altimeter was governed by a timer, not by any calculation of altitude. If Beagle had plummeted more quickly than planned, it could have hit the ground at several hundred kilometers an hour long before the airbags and even the main parachute had opened.

At a public post-mortem of the Beagle mission, held at The Royal Society, London, in March 2004, Mark Sims and Colin Pillinger appeared to accept the thinner-atmosphere hypothesis as the most likely explanation. Data from SPICAM, the atmosphere spectrometer aboard Mars Express, had suggested that at high altitudes parts of the Martian atmosphere were three times as thin as the designers of Beagle 2 had planned for. Even if the main parachute had deployed, the RAT may never have

The triple-airbag system used to cushion Beagle 2's descent failed once on testing. It remained one of the weakest links in the entire descent system.

switched in before the spacecraft hit the ground, at 56 kph (35 mph), uncushioned by airbags.

Knowing what happened to Beagle would be much easier if the project team could see it – or its remains – on the Martian surface. Surprisingly, this may be possible. Mike Malin gave up much of his time to help search for Beagle in early 2004, and some of the images returned by the MOC from the center of the Beagle 2 landing ellipse appear to show some intriguing light-tinted pixels that may or may not be the lander. Another intriguing picture only released a month after the search for a live Beagle 2 was officially called off shows Beagle in space, seconds after it separated from Mars Express, on 19 December. A small 'anomalous object' can be seen in the photo, together with a strange white patch on Beagle itself. Could a piece of Mars Express, maybe one of the bolts that held the two craft together, have collided with Beagle, damaging the British probe?

Those are the practicalities of failure. What does all this say about the philosophy of such a mission? Again from the great

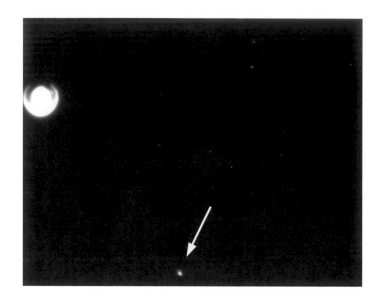

This intriguing picture, taken by the Mars Express spacecraft seconds after the ejection of Beagle 2, shows an object a few meters away from the lander. Could this have been part of the ejection mechanism that somehow became detached and collided with Beagle, damaging it in some way?

space missions. According to space historian David Harland, this was a clear case of a face being spited after the nose was cut:

> If [Pillinger] had not been so mass-constrained, he could have used a descent system that was more likely to result in probe survival. He saved a rocket – no doubt it will be used to launch a weather satellite – but I'd rather have had a 'bigger' Beagle, or three of them, on the same carrier. Spend the money to have a better than average chance of delivering the science. To cut costs to the bone and lose the probe serves no purpose.

Soon after the failure of Beagle 2 became apparent, the critics started to wade in, sometimes viciously. Jeffrey Bell, a planetologist at Hawaii University, had this to say:

> Everyone interested in Mars exploration should now take a few minutes off from looking at those fine photos of Gusev lava flow sent back by the Spirit rover. It is time to fall on our knees, face toward Memphis, and give thanks to Elvis that the British Mars lander Beagle 2 has failed. I can't think of any possible event more potentially disastrous for the future of unmanned planetary exploration than the success of this particular mission …the real disaster would have been if Beagle 2 had actually, by some stroke of dumb luck, landed on Mars and sent back some data … thank Elvis that we avoided it! Now we can all return to the boring task of designing real missions with real budgets and real prospects of success.

vantage point of hindsight, Beagle tells us that saving mass (and money) is all very well, but it is a waste of time (and money) if the mission fails. A billion dollars spent on a Mars mission that succeeds is a lot of money; a hundred million dollars spent on a Mars mission that fails is a waste of money. Beagle tells us that the mass constraints imposed by the desire to fly on ever-smaller rockets will probably result in more mission failures. Pillinger talks scornfully of the old 'battleship' missions, but the harsh reality is that the old battleships usually worked. It seems incongruous, to say the least, that Europe's own heavy-lifter, the Ariane 5, could not have been used to send a bigger, better Mars Express equipped with either a bigger Beagle 2 or a brace of Beagles into space. Instead, a hundred million euros were saved by using the relatively puny Soyuz-Fregat. But millions of euros ended up being wasted because Beagle never ended up doing any science. Could Pillinger have pushed for a bigger Beagle? Would the ESA mission masters have refused had it been a few kilos over the limit? Perhaps not. We will never know.

To the critics of faster, better, cheaper, the Beagle failure once again illustrates the dangers of having cost drive the design of

Every mission failure provokes a similar backlash. Yet it is possible to argue that much good came out of Beagle, whose legacy will be the triumph of its engineering. Before Beagle, robotic space probes were usually boxes into which instruments were crammed. Beagle did not have the weight or space for this luxury. The instruments were designed and built to fit in and around Beagle's constrained inner surfaces. Beagle 2 had curved circuit boards, bits of equipment serving functional and

'Now we can all return to the boring task of designing real missions with real budgets and real prospects of success'

Jeffrey Bell

structural purposes. This engineering triumph must not be wasted. There is no reason why some of the instruments on Beagle could not be replicated and attached to other Mars landers in the future. And there is no reason why Beagle cannot be rebuilt – maybe with a landing system with some built-in redundancy. Unfortunately, with no telemetry from the descent, it is unlikely that we will ever find out what happened to Beagle. And because of that, nobody is going to come up with even the relatively modest funding it would take to build a successor. 'Beagle 3', if such a creature is ever built, will probably incorporate some elements from the original, but the descent package will certainly have to be completely rethought.

The Beagle project was important philosophically. It was brave, and bravery is rare in the ultra-conservative world of astroengineering. Beagle was a new way of doing space, and offered an alternative to the standard NASA approach. It pushed the envelope in ways that would have been inconceivable at JPL, and that was worth doing. NASA would probably never send a probe out with only 50–50 odds of success, but sometimes you have to take risks to make progress. If it had worked, no one would have been muttering about single-point failures or too few airbags: we would all be marveling at this triumph of ingenuity. Beagle got Britain – and indeed much of the world – talking about space again. Pillinger is a PR genius, and everything he did, every step of the way, helped breathe some life into Britain's

space program, a space program that despite that country's wealth and long heritage of engineering excellence is now smaller than Belgium's. As far as the public was concerned, Britain had gone from nothing to a mission to Mars in one fell swoop, which in itself was an achievement. Beagle was put together not just by engineers working in expensive machine shops, but in their workshops at home with the most unlikely of starting materials – Beagle's drill, for instance, was a modified version of what is found in the dentist's surgery. The way Beagle was financed was also novel, and could point the way forward for other missions in the future.

Pillinger had little success in raising sponsorship from companies, but his enthusiasm and energy were enough to persuade dozens to work around the clock, often charging nothing for their services, to get Beagle off the ground. The Beagle project also enlisted the help of anyone and everyone, from rock stars to artists (Damian Hirst supplied the spot painting that was to be used to calibrate the onboard imagers). Journalists were briefed and befriended, and in the end Pillinger's project became famous, as it deserved to be. It was all terribly cheesy, but it worked, insofar as there was a project at all. If Beagle had succeeded, we would all be celebrating a very British coup. In the end, the Mars jinx, ghoul, call it what you will, struck again. In the end, Colin Pillinger and his team needed more luck than the Fates were willing to grant him. Which was a terrible shame.

Chapter 11
The Armada Arrives

'Where there is water, there may have been life'

Steve Squyres (2004)

The year 2003 ended badly for Mars exploration. There was the loss of Beagle 2, and of the Japanese Nozomi orbiter, which suffered a series of guidance malfunctions during its long voyage and in the months running up to its arrival at around the same time as Mars Express/Beagle. Any loss, even of a small and relatively cheap spacecraft, is a setback for the entire planetary community. But hard on the heels of Beagle 2's failure came three glorious successes – successes that in a few months have made Mars the star of the solar system.

Mars Express, the European orbiter (and Beagle mother ship) was put successfully into orbit around the Red Planet before the new year. By January, the High-Resolution Stereo Camera (HRSC) on board had started to send back some impressive pictures of some of Mars's more photogenic landscapes. And so have the two Mars Exploration Rovers, Spirit and Opportunity, but this time from the surface. These rovers were sent to examine two locations on Mars that may hold the key to us being able to say once and for all whether the planet really was once warm and wet. After a slightly shaky start – with, according to some sources, a hairy airbag landing for Spirit – the Mars Ghoul has been safely kept at bay and the science is pouring in. At time of writing, the dust has in no way settled on these missions. What follows is no more than a preliminary snapshot of what hopefully will be months (years in the case of Mars Express) of data. These three probes are the first in the new armada of Mars ships, an armada that will see our view of Mars transformed in the next decade or so.

MARS EXPRESS: EUROPE GOES TO ANOTHER PLANET

The sad story of Beagle 2 has been told in Chapter 10. What follows is an account of the orbiter, a large and complex (although cheap) spacecraft that is currently – and very successfully – probing Mars from orbit. Launched from Baikonur atop a Soyuz-Fregat on 2 June 2003, the 1.2-tonne

(*Left*) Mars Odyssey has lived in the shadow of its arguably more glamorous sister probe, Mars Global Surveyor. But the discoveries it has made, perhaps most notably (to date) of the subterranean water ice, have earned it a place in the annals of solar system exploration.

(*Opposite*) This magnificent image shows part of the Valles Marineris complex in unprecedented detail, courtesy of the high-resolution camera carried on Mars Express. It was one of the first images to be released by the Mars Express team.

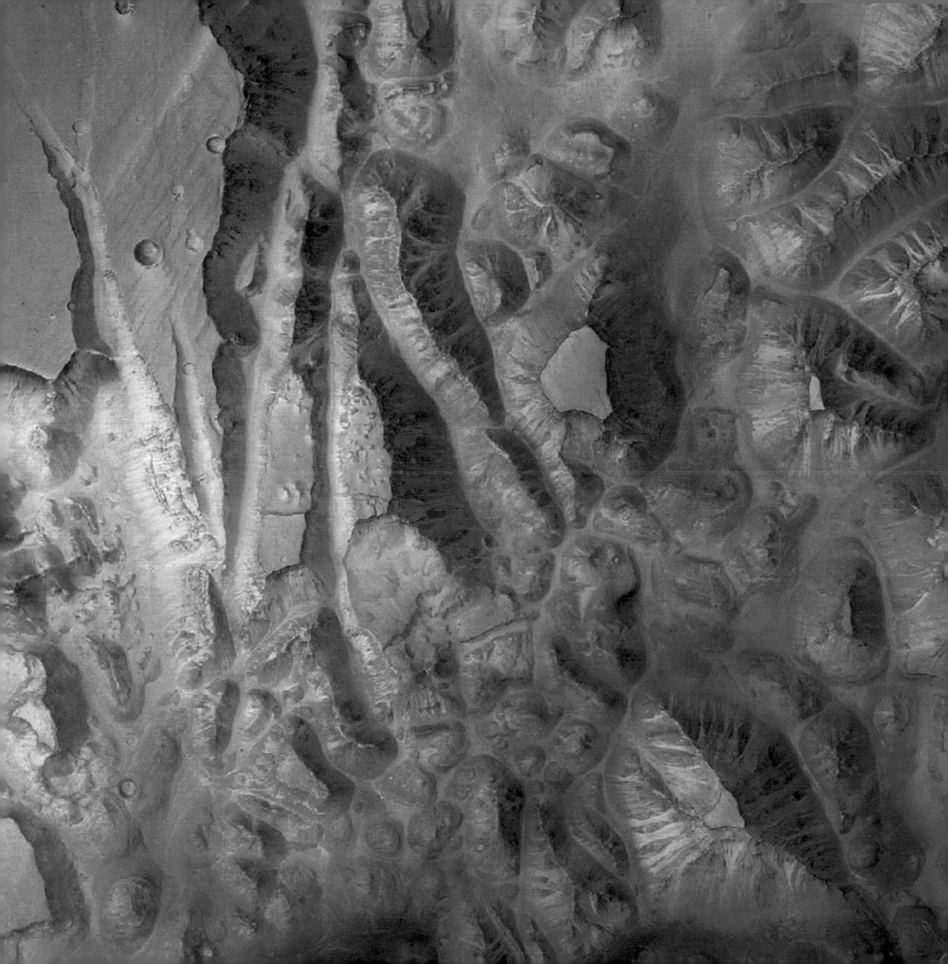

Mars Express is the lowest-cost mission to Mars so far, and, in the corporate jargon of ESA, 'is seen as a pilot project for new methods of funding and working'. It was named Mars Express because of the speed and efficiency with which the project came together. The spacecraft made extensive use of off-the-shelf components and, quite remarkably, the proposal and evaluation phase of the mission began as late as 1998, and the contract to build the thing (awarded to Anglo-French aerospace consortium EADS-Astrium) was signed only in 1999. The main cost savings came from extensive recycling – the reuse of equipment designed for the 2004 Rosetta Mission, and a number of instruments identical to those aboard Russia's lost Mars 96 probe.

The spacecraft, built mostly in Toulouse and operated from Mission Control in Darmstadt, Germany, carried seven instruments to Mars (eight, if you include the Beagle lander). Its mission is to look at the planet through a new set of eyes, concentrating on the search for water, both on the surface and at depth. Specifically, Mars Express is carrying out 'high-resolution photo-geology' – mapping Mars from space. Most of the planet will be surveyed at a resolution of 10 m (33 ft) per pixel, but selected areas will be photographed at 2 m (6.5 ft) per pixel – the highest resolution to date. The spacecraft has also been probing the surface of Mars – down to depths of a kilometer or more. If ice – or liquid water – is there, Mars Express was designed to find it. It has been carrying out its objectives as it circles Mars on a highly eccentric orbit which takes the spacecraft out to a distance of more than 11,000 km (7,000 miles), when information is sent back to Earth, and in close to 258 km (160 miles), when the main scientific observations are performed.

There are seven instruments on board. The High-Resolution Stereo Camera (HRSC) is an identical twin of the camera carried aboard the ill-fated Mars 96. The Energetic Neutron Atoms Analyzer (Aspera) studies the interaction of the solar wind with the Martian atmosphere, and the Planetary Fourier Spectrometer

(PFS) studies the atmospheric composition and circulation. A spectrometer called Omega, which operates in the visible and infrared, carries out mineralogical mapping, while the subsurface sounding radar altimeter (Marsis) probes for underground water – both solid and liquid. The Radio Science Experiment probes the internal structure of Mars, and finally the Ultraviolet and Infrared Atmospheric Spectrometer (Spicam) analyzes the composition of the atmosphere. This is an impressive suite of instruments, and Mars Express will be able to do things that no probe has done before, most notably to probe the planet for underground water

Mars Express entered orbit on Christmas Day 2003, coinciding with the fatal plunge of Beagle 2, which had separated less than a week earlier. The first result was released publicly on 19 January 2004, a spectacular three-dimensional view of part of the Martian Grand Canyon, Valles Marineris. The image was stunning, a surreal abstraction of reds and ochers. Then, on 23 January, the Omega spectrometer confirmed the presence of both water ice and frozen carbon dioxide in the Martian south polar cap. Rather puzzlingly, several media outlets gave this story the spin that this was the first time water that had been discovered on Mars. That wasn't so, of course. The Viking orbiters, by measuring the temperature of the ice caps, had shown that a large component of the northern cap in particular must consist of H_2O, as it is simply too warm in summer for dry ice to remain frozen. More recently, towards the beginning of 2002, Mars Odyssey's gamma-ray spectrometer detected large quantities of water ice all over Mars, even at low latitudes. This wasn't the only thing about Mars Express's early findings that raised a few eyebrows. There also appeared to be something strange about these first pictures.

If there is one thing that the Face on Mars fiasco should have told anyone operating a space probe, it is to be careful with the pictures you release to the public, otherwise it is all too easy for some people to get the wrong impression. While the 'Grand Canyon' picture showed Mars very much in its true colors, over the coming days photographs started to come rolling in that most definitely did not. One image in particular caught everyone's eye – an overhead shot of Reull Vallis, a channel several kilometers wide in the southern hemisphere, east of the Hellas basin. Slightly controversially, the ESA press release described the channel as 'being carved by running water'. But far more controversially, the photograph showed

(*Opposite*) An artist's impression of liquid water lurking hundreds of metres under the Martian surface. It is already known that large deposits of water ice lie just beneath the surface at low latitudes, and many scientists believe that at greater depths, where liquid water would be stable (and kept warm perhaps by geothermal heat), vast reservoirs of 'blue gold' may be found. If underground aquifers like this really do exist, Mars Express has a good chance of finding them with its subsurface radar instrument. The implications for human exploration and the possible existence of Martian life would be profound.

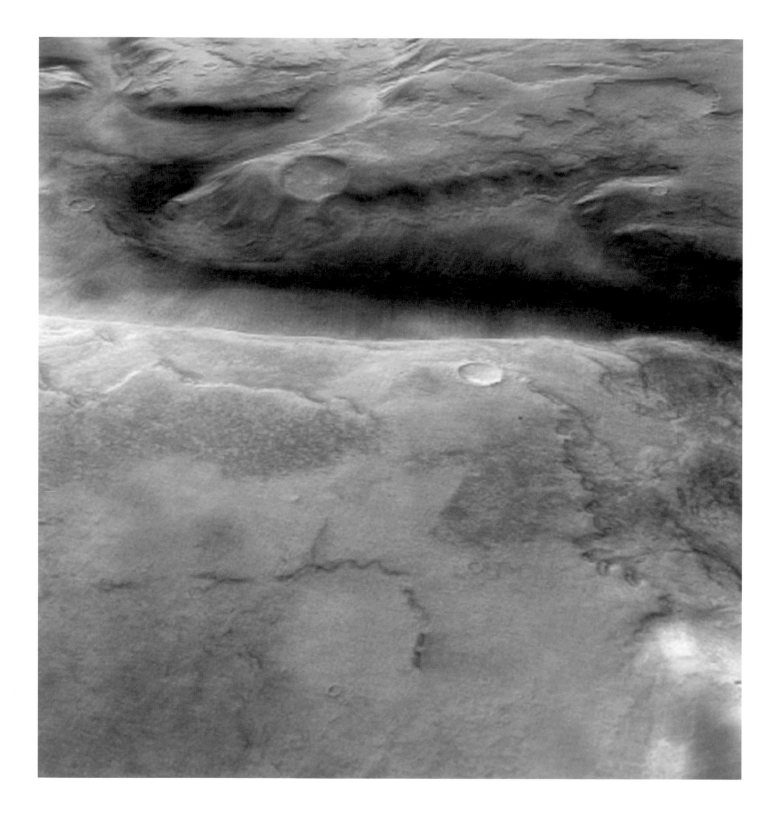

(Above) Surely one of the most spectacular images of Mars to date. This picture was taken by the High Resolution Stereo Camera (HRSC) on board ESA's Mars Express orbiter on 19 January 2004. It shows a three-dimensional oblique view of the summit caldera of Albor Tholus, a volcano in the Elysium region, with the vertical scale enhanced by a factor of three. The caldera is 30 km (20 miles) across and 3 km (2 miles) deep; the whole has a diameter of 160 km (100 miles) and a height of 4.5 km (3 miles). This is geologically interesting, since the depth of the caldera approaches the height of the volcano, which is unusual on Earth. On the far left rim of the caldera, a bright 'dust fall' seems to flow from the surrounding plateau into the caldera.

(Opposite) This image by the Mars Express orbiter has been processed to give a three-dimensional quality. It shows a large channel, the Reull Vallis, just east of the Hellas Basin. Problems with calibrating some of the early images led to the coloration being rather unrealistic, and there was criticism from some quarters that garish images like this were creating a false impression.

the floor of the valley in a dark blue color. Another image showed Gusev crater, the landing site of the NASA Spirit rover. Rather than the reds, browns, and grays of Mars, we were treated top a dash of vivid green. A spectacular picture – probably the most incredible yet, and one surely destined for posters – shows the caldera at the summit of Albor Tholus, an extinct volcano in Elysium. The shot is remarkable because it shows what appears to be a dust 'waterfall' cascading into the caldera from the summit rim. These pictures are wonderful, but are they showing us the real Mars?

It wasn't long – only a day or so – before the conspiracy theorists thought they had smelled a rat. 'Is this why NASA has been lying about the true colors of Mars?' asked the Enterprise Mission website, just after the European Gusev image was released. On the website was a picture showing Gusev in its 'false' (i.e. true) 'NASA' colors – red and brown – alongside an image of the crater in its 'true' (i.e. false) green-enhanced hues, courtesy of Mars Express. This was not the first time that the Mars conspiracists have alleged the airbrushing of the color green – surely indicating a lush carpet of vegetation – from Martian history. And other people wondered whether this was a case of a bit of drama being added to the images. After all, greens and blues make for more attractive pictures than grays and browns. Of course, there is no cover-up, but the strange colors do need explaining.

Jan-Peter Muller is a senior member of the HRSC team, and developed the software that renders the Mars Express images in 3D. He says there had been two problems with getting the cameras properly calibrated. First, at the beginning of 2004 there was a lot of dust in the Martian atmosphere, which had a distorting effect. Secondly, the images were released before the color balances had been correctly set. 'It was thought better to get something out straight away, even if it was not quite right, than to have a big delay', he says.

That first image released, the one of the Grand Canyon, was a very vivid red, which is extremely close to the real color. The blue in the Reul Vallis is down to the fact that one of the channels had very poor contrast and the grayish colors turned to blue. That is, basically, a mistake, and these things happen. The images will become more color-coordinated – there has certainly been no attempt to jazz up Mars here.

The pictures coming from Mars at time of writing might as well be of a different planet to that photographed by the early Mariners. This Mars Express view shows the western flank of Olympus Mons. This cliff is more than 7,000 m (23,000 ft) high.

AMERICA ARRIVES: SPIRIT AND OPPORTUNITY

It has not been a happy time recently for NASA's efforts to explore Mars. The losses of Mars Climate Orbiter and Mars Polar Lander in 1999 were a crushing blow to JPL. With the twin Mars Exploration Rovers of 2004, NASA had to get it right. In the end, the missions were more successful than anyone had dared to hope. The Mars Exploration Rovers (MERs) get down safely, bouncing to a halt after their descent was cushioned by no fewer than twenty-four airbags each, and both spacecraft have produced huge amounts of data.

The MERs are mobile robotic field geologists. While Sojourner was able to carry out limited traverses and examine in situ several of the more interesting rocks within a 50 m (160 ft) radius of the Pathfinder landing station, the MER brief was far more ambitious – to explore a large area of Martian terrain, ranging several hundred meters from the landing sites. Two sites were selected because they were thought to be of particular geological interest – a large, partially buried crater which looked as if it might be the site of a lake at some time

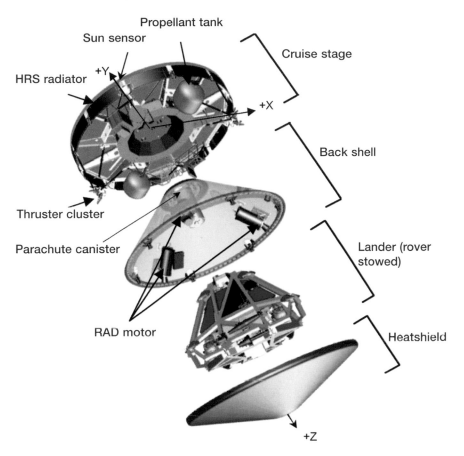

The twin landers had a complex and elegant descent and landing system which made use of three technologies – a parachute, guidance rockets, and airbags.

NASA's 'follow the water' strategy is what is currently driving Mars exploration. The findings of the Spirit and Opportunity rovers seem to show that the strategy is justified, but some scientists caution that focusing the exploration of Mars on finding water risks building public hopes up too high that life will be found.

in the distant past, and a small patch of the Martian surface known from orbital spectroscopic observations to be rich in the iron mineral hematite – and also, perhaps, either the site of geothermal activity or what might once have been a large body of standing water. The choice of both sites was governed by the JPL's follow-the-water philosophy. To carry out their task, the twin MERs were equipped with an impressive suite of instruments – the scientific package being collectively known as Athena – including spectrometers, no fewer than nine cameras each, a microscope, and a rock drill. The specific mission objective was to see whether the geology at the two landing sites suggested that there was once freestanding water on these parts of Mars.

It is also worth pointing out what the MERs were *not* designed to do. For a start, they were equipped with no instruments capable of detecting extant or fossil life directly – only to seek clues to an environment that could have supported life (or possibly still does). It is feasible, of course, that rocks containing visible or microscopic fossils could be discovered, but this was always a very long shot (extant, non-microbial life could also be discoverable in theory of course). It was always the much-disputed contention of the Beagle team that their spacecraft was superior in this respect, having instruments on board designed specifically to look for the chemicals – notably methane – that are the signatures of microbial activity.

The first of the twin MERs to arrive was Spirit, which bounced to a halt on the floor of the Gusev Crater 3 January 2004. In contrast to the lost Polar Lander, the MERs provided telemetry of a kind – not a stream of data, but instead a series of tones, each of which indicated that a specific part of the descent system had been activated, so that if something went wrong then at least the engineers would know where to begin looking for a problem. When the lander signaled that all was well, there was sustained and prolonged cheering at JPL mission control, led by the charismatic science team leader for the missions, Steve Squyres. Everyone was delighted that the mission was a success, but also relieved: if Spirit had failed through some fundamental design fault, then Opportunity, now closing in on Mars fast, could be in trouble too. Both probes had been designed and built extremely rapidly – not cheap (about $400 million apiece) but certainly fast. The question was, were they going to be better? No one knew whether JPL's Mars exploration program could survive the loss of another two probes on top of the humiliations of 1999, and the Columbia Shuttle tragedy. If both the MERs failed, serious questions might even be raised about the future of NASA itself. Fortunately they didn't. The cheering resumed three hours later, when the first images were sent back.

The landing sites for the MERs had been chosen carefully. It is not appreciated by many how difficult it is to choose a place to put a lander down on Mars. To the layman, it might seem obvious that the best places to go would be the most dramatic – the floor of the Valles Marineris, for example, or the middle of the caldera atop Olympus Mons. But sadly, although such sites would indeed be of tremendous scientific interest, and could provide some spectacular vistas for the probe's cameras, they are simply too dangerous. Spectacular landscapes tend to mean steep slopes, deep craters, and other hazards which could smash, disable, or render a precious spacecraft incommunicado. The trick was to find flat places which looked as if they would tell us about water.

Two years of research and effort went into selecting the MER sites, and the views of dozens of scientists and engineers were taken into account. The main constraints were engineering – specifically, the rovers had to land at points significantly below Martian 'sea level'. At a relatively low altitude the descending probe travels through more – and thicker – atmosphere, allowing the parachute to slow it down more effectively and giving more time for the radar altimeter to measure the closing velocity and ready the mechanisms for inflating the airbags and firing the rockets that guide the probe on its final descent. Another constraint was the need for power. The rover twins were solar-powered, so polar landing sites were out of the question. Both landing sites had to be within 15° of the Martian equator. Also, the two rovers had to be several thousand kilometers apart, otherwise it would have been problematic to relay the signals back to Earth via the Mars orbiters, as the two signals could interfere.

Those were the broad-brush considerations. Now the team had to go looking for interesting places. Steep slopes were ruled out not only because they make landing hazardous, but also because the rovers would not be able to operate effectively on terrain that is too rugged. It was decided that slopes on the landing site should be less than 2° over a 1 km (0.6 mile) range, and less than 5° over a 5 m (15 ft) range. The airbags were designed to protect the landers from being smashed to bits by rocks up to half a meter (20 inches) in diameter. The sits chosen also had to be load-bearing and not too dusty (ruling out sand dunes).

Nearly two hundred potential landing sites were found that met all the criteria and which looked interesting. There were areas showing signs of water erosion, ridged plains, and places that looked as though they had been shaped by volcanic action. Slowly the choice was narrowed down to twenty-five, including one actually inside the Valles Marineris complex, and several possible crater lakes. Agreement was eventually reached on four prime sites and two backups.

It looks a mess of plumbing, rigging, and electrics, but it all kept Spirit in one piece as the lander made its perilous descent through the Martian atmosphere.

The final shortlist of sites were all possible 'wet spots'. All showed evidence, according to their advocates, of surface processes involving water and appeared to meet the principle science objectives of the MER missions, namely to determine the aqueous, climatic, and geological histories of sites on Mars where evidence 'may have been favorable to the preservation of evidence of past or present life'. If the landers were to find little or no evidence of water, then the likelihood of there ever

having been Martian life would fall dramatically. But if strong evidence of water were found – rocks which have been weathered or eroded in a watery environment, or the minerals which are known to be deposited in an aqueous environment – then the warm wet hypothesis was probably correct. Their role, therefore, was to finally resolve the basic problem of Mars: were all those features that look as though they were made by water *really* made by water?

Mars Odyssey had shown that at Meridiani, an equatorial plain a few hundred kilometers east of Marineris, there were large deposits of coarse-grained gray hematite, an iron oxide mineral that could have formed by precipitation from water,

This spectacular picture was taken by Mars Express in early 2004 and shows a mesa to the east of the Valles Marineris. The large crater in the background has a diameter of 7.5 km (4.75 miles) and a depth of 800 m (2,600 ft). This was one of the landing sites under consideration for the two Mars Exploration Rovers.

The dusty roads of Mars . . . Spirit's tire tracks left in the sand.

or as a hydrothermal deposit. The crater Gusev has been interpreted by some researchers as an ancient lakebed, with interior sediments deposited in standing water. Gusev, which is about 160 km (100 miles) across, was created by an impact early in Mars's history – about 4 billion years ago. A large channel called Ma'dim Vallis, runs into the crater from the southern highlands. The channel is nearly 850 km (530 miles) long and some 40 km (25 miles) wide, and looks as though it carried water into the crater for a long time. It is a classic example of what looks like an ancient riverbed. Spirit landed in the middle of Gusev on 4 at 11.35 p.m. (Eastern Standard Time) on 3 January; its twin rolled to a halt in Meridiani three weeks later.

The twin Mars rovers are the latest members of a family of small, lightweight vehicles that have been developed by NASA and its partners in academia over the last two decades. The MER can trace its genes back to a prototype design-study known as Rocky 1, developed by JPL in the late 1980s. Rocky 3 was an improved model which also contained a robotic arm and

sensors. In trials it could navigate by itself to an interesting location, moving around any obstacles in its way, and then scoop up a soil sample. It could then return to base – the stationary lander part of the spacecraft – by using an infrared beacon as a homing device. The next generation, Rocky 4, was a flight prototype version of Rocky 3: its chassis was made light and strong enough to withstand the stresses of launch and landing. A key additional instrument was a device to chip away at rocks. A successful June 1992 test of Rocky 4 paved the way for a rover to be included on the Pathfinder mission.

JPL scientists have been testing their Mars rovers in sandpits and deserts for nearly a decade. Back in 2000, a robot called FIDO found itself confused – it had run into a bush out in the Nevada desert. Its internal software was simply unable to cope with an object it would not expect to see on Mars. It has been the continuing exponential growth in computing power that has effectively supercharged probe design. The MER rovers contained 'brains' that are about as powerful as a 2001 laptop computer. This doesn't sound very impressive, but in space terms it is cutting-edge: 128 MB of RAM is about 200 times as much computer memory as the Apollo spacecraft had. Most computer systems fitted to spacecraft –

This picture, taken by Spirit some four months after landing, shows the first close-up look at the Columbia Hills, about 3 km (2 miles) from the spacecraft's landing site in the crater Gusev.

from the Apollo landers to the Voyagers and other deep-space probes – have been hideously out of date by the time they fly. This is because, quite rightly, NASA considers reliability to be more important than speed.

The MERs are five times taller, nearly eighteen times heavier, and, by being taller, they can see six times farther than their predecessor. Sojourner had just three cameras; of the nine carried aboard each MER, five are monochrome and four color, including two mounted on a mast more than a meter and a half (5 ft) tall known as the Pancam Mast Assembly, which also acts a periscope for one of the spectrometers. These twin cameras, mounted on their face-like mounting structure, gives the MER an appealing, comic-book appearance. When it comes to roving, the MERs are in a different league to Sojourner. The little pioneer was capable of moving only 5 m (15 ft) a day, and never traveled more than a 100 m (330 ft) from where it landed. Both the MERs were designed to travel up to 100 m a day, and in theory could each trundle a total distance of several kilometers before the missions ended.

Each of the MERs' six wheels has its own independent motor. The two front and two rear wheels are steered individually, allowing the vehicle to turn on the spot. The sophis-

ticated steering mechanism gives the rovers the handling and road-holding ability of a Formula One car combined with a Humvee (although they are of course considerably slower, even than the latter). The suspension units on which the wheels are mounted were developed for Sojourner. The whole suspension–steering system allows the MERs to climb over obstacles larger than the diameter of their wheels while keeping all the wheels on the ground. And the rovers can tilt up to 45° in any direction without tipping over.

The other instruments aboard the MERs are also far superior to what was carried on Sojourner. Where the earlier rover had a single spectrometer capable of detecting alpha-particles, protons, and X-rays, the MERs each have three: one similar to that on Sojourner, a Mössbauer spectrometer specifically to detect iron compounds, and a mini-thermal emission spectrometer (Mini-TES), which is a type of Michelson interferometer. Mössbauer spectrometry relies on a phenomenon called nuclear resonance absorption and was developed in the 1950s as a way of detecting the presence and relative abundance of particular elements with a fine degree of accuracy. The Mini-TES probes the mineralogy of nearby soils and rocks, by analyzing their light in the near-infrared part of the spectrum. It is there specifically to search for minerals that were formed by the action of water, such as carbonates and clays.

We know from the previous probes to land there that Mars's surface is both dusty and very old. That has implications for the study of material on the surface. First, pointing a spectrometer at a rock may not be enough to tell you what it is made of if it is coated in dust, for the dust may have blown there from hundreds or even thousands of kilometers away, in which case you will get a false reading. Secondly, aeons of exposure to sunlight, wind, cosmic rays, and moisture will inevitably have changed the surface composition of any material exposed to the Martian atmosphere. On Earth, the combined action of the elements with solar radiation leaves desert rocks coated with a varnish-like layer called desert glaze, which is chemically quite distinct from the material inside. Neither the Viking landers nor Sojourner were equipped to break open or scrape rocks to get at an unadulterated sample. The MERs solve this problem by positioning a device called the Rock Abrasion Tool (RAT) against a rock. The RAT has a grinding wheel to remove dust and weathered rock, exposing fresh rock underneath; it can expose an area of nearly 50 mm (2 inches) in diameter and grind to a depth of about 5 mm (0.2 inches).

While the main cameras are held aloft by the Pancam assembly, most of the instruments that need to get down and dirty with rocks and regolith are bolted to the Instrument Deployment Device, a robotic arm complete with shoulder, elbow, and wrist. This arm allows the Mössbauer, the RAT, and the onboard microscope (which can focus down to around 0.1 mm, or 1/250 inch) to get close-up views of objects on or near the ground.

Ideally, all rovers would be powered by atomic batteries. NASA's most successful long-duration missions have all carried Radioisotope Thermal Generators (RTGs), which can provide high levels of power for months or even years. For missions to the outer solar system, where sunlight is in short supply, RTGs are essential; the Cassini probe, due to arrive at Saturn in late 2004, would receive there just 1 percent of the sunlight that falls on Earth – nowhere near enough to generate sufficient electricity to power its instruments. Mars is much closer to the Sun than Saturn, but solar power is not ideal. The sunlight is till weaker than on Earth (although there are fewer impediments to its progress down to the surface) and Mars is very cold, especially at night. Under such conditions, the batteries that must be carried to store power would not fare well. In addition, a coating of dust builds up slowly on anything exposed to the Martian atmosphere; solar panels, which are big, flat surfaces that must be held more or less horizontal to the ground to maximize their efficiency, get dusty very quickly

on Mars, and after a few months their efficiency starts to dwindle rapidly.

But nuclear power is not popular at the moment. So the MERs are equipped with solar panels, and they have to supply power for everything – the wheel motors, all the science instruments, and so on – and keep the electronics warm. The panels are designed to generate around 140 watts of power for up to four hours a day at the beginning of the mission. By the end of the mission, though, the combination of dust, weakening Martian sunlight as the seasons change, and the inevitable degradation of the batteries, will have reduced this to just 50 watts – barely enough to light a room, let alone keep a hefty mobile robot fed.

The importance of communication when it comes to unmanned probes cannot be overstated. The Galileo mission was nearly lost when its high-gain antenna failed to unfurl properly *en route* to Jupiter; and the losses of Beagle 2 and Mars Polar Lander, while certainly not a direct result of communication problems *per se*, were exacerbated by a lack of telemetry from the descending probes which might have allowed mission control to find out what had gone wrong. The dependence of Beagle 2 and, to a lesser extent, the MERs on existing Mars orbiters to act as relay stations to Earth highlights this potentially weak link. The problem of communicating with Mars probes on the surface will be solved only when a network of communication satellites is placed in orbit around the planet, allowing 24-hours-a-day (or 24-and-a-bit hours a day) talking to home.

In the meantime, a mixture of direct communication with Earth and relay signaling must be used. The rovers each carry three antennas. The High-Gain Antenna can be pointed directly at Earth to communicate with the giant radio antennas of the Deep Space Network. The UHF Relay Antenna transmits signals between the rover and orbiting spacecraft – Mars Global Surveyor, Mars Express, and Mars Odyssey. There is also a small UHF low-gain pole antenna, which will be used to transmit information about the rover's mechanical health and operating systems back to JPL.

Following the success of Pathfinder, an airbag system was decided upon for the MERs missions. An aeroshell and parachute achieved the initial deceleration; retro-rockets were then fired to further slow the lander's speed of descent and also to correct for any lateral drift (as Spirit was landing,

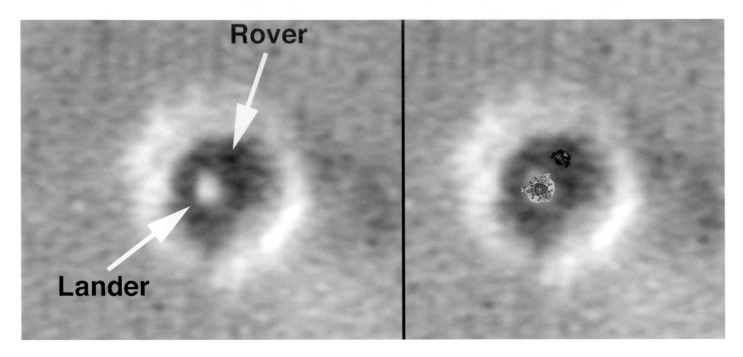

Rover

Lander

These images, taken from orbit by the Mars Orbital Camera aboard Mars Global Surveyor, show the rover Opportunity and its lander in the Eagle crater. In the right image, computer-generated scale images of the rover and lander have been superimposed.

significant course corrections had to be made as windspeeds were higher than had been anticipated). Finally, twenty-four airbags were inflated to cushion the lander on its impact with the surface, and the whole package to bounce and across the surface for nearly a kilometer before rolling to a stop. The airbags were then deflated and detached, and the lander's body segments opened out like the petals of a flower.

The key part of the landing system was the airbags, a recent and imaginative solution to reducing weight on landers. No parachute alone can slow a spacecraft to a safe touchdown speed in the Martian atmosphere, so some sort of final descent system is needed. The airbags used for the MERs were the same type that Mars Pathfinder used in 1997. Airbags must be strong enough to cushion the spacecraft if it lands on rocks or rough terrain and allow it to bounce across the surface at speeds of up to 100 kph (60 mph) after landing. A mechanism is needed to inflate the airbags seconds before touchdown, and deflated them once the lander is on the ground and has come to a halt.

Each rover used four airbags with six lobes each, which

are all connected. Once the bags were inflated, the descending spacecraft must have looked like giant bunches of grapes falling from the sky. The landers were contained in a bivalve-like arrangement consisting of a cork-faced heat shield and a white backshell, which were fitted with three large rocket motors called RAD rockets (Rocket Assisted Descent), each providing about a tonne of force for over 2 seconds. Happily, every component of this complex system worked perfectly.

The MERs have had a difficult task to perform on Mars that pushes the limits of robotic exploration technology. Each rover must be instructed by mission control many hours in advance – because of the distance between Earth and Mars, nothing like real-time communication and feedback is possible. Although the twin rovers are a huge advance on Sojourner, they are limited to about 100 m (330 ft) of travel a day. The complicated hazard-avoidance software on board enables the machines to make autonomous decisions, principally avoiding rocks and potholes. But the whole process can be painfully slow: it can take a week from when the rover's camera spots an interesting rock or area of sand to when it is actually probing it with its instruments and sending back data. Finally, there is the dust – the fine

(*Opposite*) Splashdown! This surreal image, a tangle of fabric, aluminum, jumbled solar panels and electronics, is in fact the Opportunity lander snapped by its own camera, held aloft on the Pancam assembly.

orange powder that will probably prove to be the bane of astronauts' lives if we do get around to going to Mars in person. The missions were supposed to last just 90 days, but early on it became clear that this was a conservative estimate of how long the solar panels could keep functioning, and by March the rovers' predicted lifespans had been increased to eight months.

MAGIC CARPETS, BEDROCK AND BLUEBERRIES

As the pictures started to roll in, it became clear that there was much about the Gusev site that was familiar, but also much that was strange. The vista seen by the camera aboard Spirit showed a greater range of relief than at any other previous landing site, and included some substantial hills, named the East Hills, 100 m (350 ft) high and about 3 km (2 miles) away on the horizon, and (just within driving range) a more substantial hill, again poking up over the horizon. The foreground was a rock-strewn plain, similar in

appearance to the rocky plain seen by both Viking landers and Pathfinder. A small, apparently sand-filled depression a few meters from the lander was soon christened Sleepy Hollow.

Almost immediately there was a flurry of excitement when traces of carbonate minerals showed up in the rover's first survey of the site with its infrared sensing instrument. Gusev had been chosen as a landing site because it may have once been a lakebed, and the scientists were excited by what appeared to be limestone. The pictures also showed some very odd soil around Spirit, which had an almost mudlike consistency. This soil, quite unlike any seen before on Mars, appeared to be 'stuck' together, folding and warping like a mat where it had been scrunched up by the lander's airbags. It was dubbed the 'magic carpet'.

But what about those carbonates? The initial excitement soon tailed off. Their low concentration suggested, in fact, that these were not formed under water, as had been hoped, but instead were simply the result of aeons of chemical weathering by the trace amount of water vapor in the Martian atmosphere. It was part of the dusty rind on the rocks. In fact, a low-level carbonate signature (around 1–2 percent) is found all over Mars. The rocks strewn around Spirit's landing site were mostly dark and smooth, and look like basalt, the dark volcanic rock that comprises the majority of the Martian crust. The smoothness may be the result of aeons of wind-blasting by fine silicate dust, or of water erosion. There was no visible layering, no cross-bedding (fine layers at an angle to the main strata), no embedded pebbles or nodules – nothing to suggest that these rocks were anything other than purely volcanic.

Spirit suffered a serious computer glitch – a breakdown in the probe's flash memory – which kept it out of action for several days. There were initial fears that Spirit had been lost, and then talk of needing weeks to fix the problem. But NASA has experience in repairing and modifying computers that are hundreds of millions of kilometers away: Galileo's onboard system was reprogrammed when it was

It has become a tradition since Pathfinder days to name just about every rock and dune seen on the surface of Mars. This true color image taken by Spirit shows 'Adirondack,' the rover's first target rock, which it trundled up to three days after emerging from its landing cocoon.

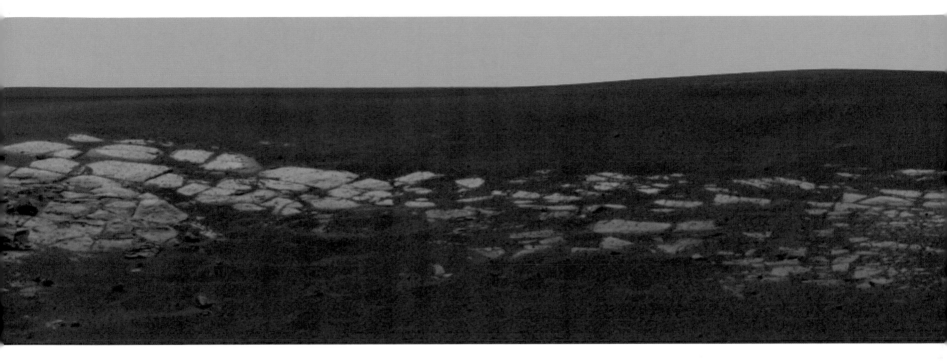

The rocks that launched a thousand journal articles. These layered rocks in the wall of the Eagle Crater, in which Opportunity found itself after bouncing to a halt, have provided more answers than many geologists dared hope. But they have also posed more questions. The picture does not give a good sense of scale – the rock face is no cliff, but about the height of a sidewalk curb.

halfway to Jupiter to take account of the loss of the probe's high-gain radio transmitter. A 'patch' was transmitted to Spirit, and it was soon back on its feet. A few days after the computer was repaired, Spirit got to work on a nearby rock that was named Adirondack, drilling several millimeters into its surface and showing that it was composed of olivine-rich basalt.

Meanwhile, on the plains of Meridiani, Opportunity had landed in a small crater just a few meters across. The vista that greeted the scientists at JPL was quite unlike any other that had been relayed from the surface of Mars. Instead of the boulder fields seen at the other four landing sites (the two Vikings, plus Pathfinder and Spirit) there was just a flat, featureless dark brown plain, which could be glimpsed over the rim of the crater, perhaps covered in a thick layer of pyro-

clastic deposits. An outcrop of white, layered rocks caused much excitement. This was the first bedrock to be seen on Mars by a lander. On Earth – and on any other planet – exposed bedrock is a window into the deep past. Other surface material – loose boulders, sand and regolith – will have been liberated from bedrock by chemical or physical weathering, and will consequently have been altered.

Thousands of tiny nodules dubbed 'blueberries' (although they are not actually blue) were found scattered all over the crater. Subsequent examination from close up revealed that they had probably been eroded out of the light-colored rock formation and were most probably composed of hematite. According to Nick Hoffman, an enthusiastic contributor to the whole Mars debate,

It's what we desperately want to understand about Mars – just what is this pale rock? Is it sediment or a volcanic ash? What is the relationship between the pale rocks and the dark dust? What were conditions like when it was deposited? Frankly, [even] if Opportunity never gets out of this crater, it has paid off.

The outcrop of bedrock was extremely important.

THE MARTIAN BUNNY

When the first high-resolution pictures were beamed back by Opportunity straight into the welcoming arms of the World Wide Web, they were immediately scrutinized by several hundred million pairs of eyes. There were the strange white rocks, a lot of very fine, reddish-brown sand, a flat horizon poking out from beyond the crater wall – and what looked like a small white rabbit sitting a few meters from the spacecraft. It had what looked like two pointy ears sticking up, and soon became the subject of some light-hearted speculation.

Initially, JPL seemed at a loss for an answer. The public and the media were fascinated, and emails pinged their way back and forth with suggestions about what this wee beastie might be. The conspiracists claimed early on that NASA had 'killed' the rabbit by promptly crushing it under Opportunity's metal wheels, but even they were at a loss to explain what the object was. The 'bunny' is a yellowish-gray object about 50 mm (2 inches) long. Possibly the first person to spot it was Jeff Johnson, a scientist at the US Geological Survey. 'What in the world is this?' he asked his colleagues. Then, as soon as it had been spotted, the rabbit disappeared. It turned out that it had blown a few meters in the light Martian breeze (and Martian breezes are extremely light indeed), suggesting that whatever it was made of was itself very light. The bunny ended up under the ramp used by the rover to roll off its landing base. The images acquired of the object as part of the panorama even showed some evidence that the 'ears' were swaying in the breeze between each shot.

So what was the Martian Bunny? Almost certainly it was a piece of the spacecraft, possibly part of the airbag system. 'Our team believes that this odd-looking feature is a piece of soft material that definitely came from our vehicle,' stated Rob Manning, lead engineer for entry, descent and landing in a JPL release on the object.

End of mystery, although no doubt the bunny will be added to the Martian hall of fame.

The bunny. This strange object, complete with 'ears' that flapped in the wind, was a source of much amusement and bemusement. It turned out to be a piece of airbag or covering from the spacecraft, rather than an unexpected specimen of large Martian fauna.

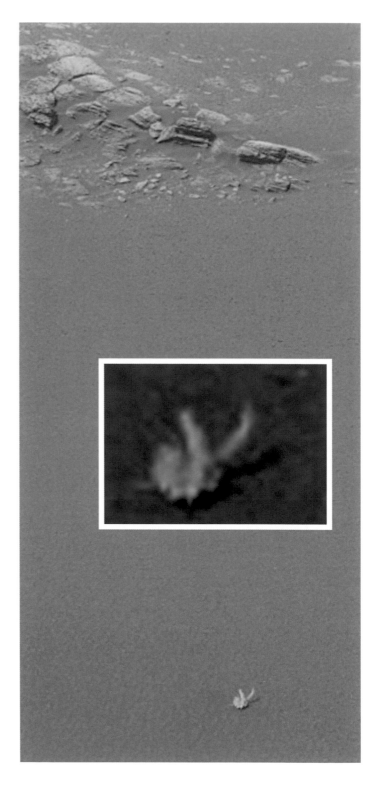

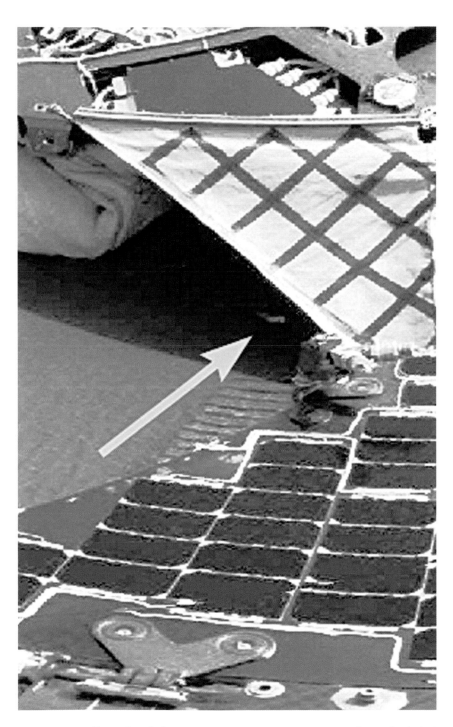

Bunny found! Jeff Johnson, a scientist with the United States Geological Survey on the Opportunity team, was assigned the job of tracking the bunny. It was eventually found, hiding under the rover's egress ramp.

TENSION MOUNTS

'We've concluded that the rocks here were once soaked in liquid water'

Steve Squyres

For about a fortnight after the arrival of the rovers, there was a constant stream of news from JPL. There were the pictures, of course, and daily press briefings at JPL's HQ in Pasadena. Some scientific results were released – the carbonate signature at Gusev – and there was much speculation about the nature of the hematite found at Meridiani and, of course, those intriguing blueberries. Then, nothing. For nearly a month, it seemed the JPL website was silent. The press briefings dwindled and became rare events. There was no hard information, no news, no data. Meanwhile, in the informal planetary sciences community, speculation started to grow that JPL had something really special up its sleeve. On 2 March, JPL called an urgent press briefing. What Squyres and his team announced was indeed truly startling. Opportunity, the rover exploring the Meridiani site, had found 'definitive' evidence that this part of Mars at least had once been sloshingly wet. 'We've concluded that the rocks here were once soaked in liquid water,' Squyres announced. There was no doubt, in the minds of the JPL scientists, that the instruments on board Opportunity had done what they were designed to do – they had followed the water and found it. It wasn't long before the 'L'-word – life – was being used.

NASA had not discovered running water on Mars. But it has discovered extremely strong evidence that the rock outcrop visible along the edge of Opportunity's crater had once been immersed in water, and may even have been deposited in water. Ironically, it was not the presence of hematite that led the JPL scientists to this conclusion. Meridiani, an Oklahoma-sized patch of Mars known from orbital studies to be rich in this iron mineral, was chosen because if the presence of hematite could be confirmed on the ground, it would be strong circumstantial evidence that this part of Mars was once under water or a hydrothermal system. But it was not hematite but that whitish layered outcrop, dubbed El Capitan, that was the geological bulls-eye as far as Squyres and his team were concerned. As Squyres said at the press conference, 'Liquid water once flowed through these rocks. It changed their texture, and it changed their chemistry.' In an accompanying statement Jim Garvin added that

NASA launched the Mars Exploration Rover mission specifically to check whether at least one part of Mars ever had a persistently wet environment that could possibly have been hospitable to life. Today we have strong evidence for an exciting answer: Yes.

What Opportunity had obtained was spectroscopic evidence of a concentration of sulfate minerals in the bedded rocks forming the nearby white outcrop. In addition, there are holes in the rocks in the shape of crystals, suggesting that minerals were deposited out of a watery solution and were then weathered away. Magnesium, iron, and bromide salts were found. The rover's Mössbauer spectrometer, which identifies

iron-bearing minerals, detected a hydrated iron sulfate mineral named jarosite. The presence of jarosite indicates that the rocks may have been steeped in an acidic lake or a hot spring.

Fortunately, there were other things about these rocks that gave valuable clues to their formation. Across the whole outcrop could be seen signs of cross-bedding, suggesting that these rocks were deposited as sediments – maybe by water, but also perhaps by wind, or even in a pyroclastic flow. The indentations, known as 'vugs', may have been home to precipitated salt or gypsum crystals. Spherical concretions – the blueberries – look like water-deposited nodules. Taken together, the case for ancient liquid water on Mars, at least at Meridiani, looks strong. Putting multiple lines of evidence together, 'it's hard to avoid the conclusion that this stuff was deposited in liquid water', said chief scientist Steve Squyres:

Now, what's the significance of this? The purpose of this mission was to go to Mars and see whether or not it once had habitable environments. We believe, at this place on Mars, for some period of time, it was a habitable environment. This was a groundwater environment; this was the kind of place that would have been suitable for life. Now that doesn't mean life was there. We don't know that. But this was a habitable place on Mars at one point in time.

After spending several weeks examining the treasure trove of rocks inside its cozy crater, Opportunity climbed out and began exploring its surroundings more widely. This picture shows the 22 m (72 ft) wide Eagle crater from the level of the surrounding plains.

Squyres continued:

Ever since Opportunity touched down on Meridiani Planum the night of January 24 and we first opened our eyes and took a look around and saw this marvelous outcrop of layered bedrock literally right in front of us, we've been trying to puzzle out what this outcrop has been trying to tell us. For the last two weeks, we've been attacking this outcrop literally with everything we have, every single piece of our payload has been brought to bear on this. Over the last couple of weeks, the puzzle pieces have been falling into place and the last puzzle piece fell into place a few days ago. We have concluded the rocks here were once soaked in liquid water.

Where there's water, there may have been life. But how sure can he be? Squyres told the BBC a few days after the 'water'

(*Right*) This false-color image taken by Opportunity shows clearly the spherical objects that are scattered everywhere around its landing site – in this case, they are strewn around the rock named Stono Mountain. The colors have been changed to exaggerate the differences between the spherules and the rock. The rover's Mössbauer spectrometer examined a patch of the tiny spherules (also dubbed blueberries, despite not being blue) and revealed them to be composed of hematite, and iron oxide mineral that was probably formed under water, or at least in damp conditions.

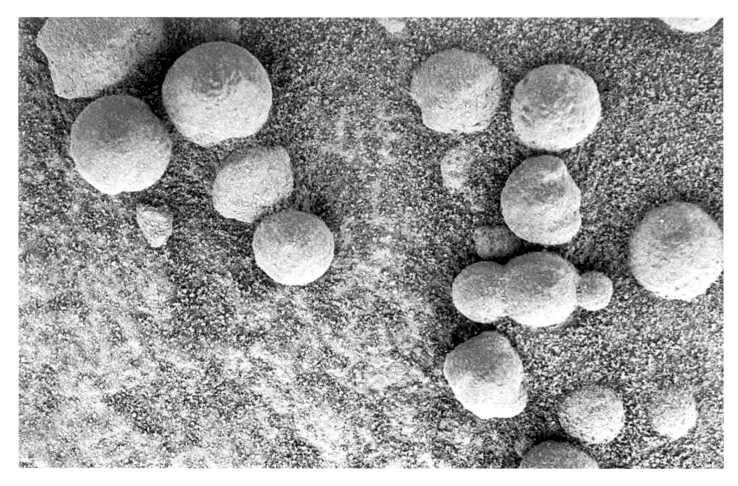

This microscopic image was taken at the outcrop region named Berry Bowl near Opportunity's landing site. It shows, close up, the intriguing spherical objects scattered all across the Meridiani plain. Of particular interest is the blueberry triplet, towards the right of the image, which indicates that these objects formed in pre-existing wet sediments. Other spherical grains that form in the air, such as impact spherules or ejected volcanic material, are unlikely to fuse along a line and form triplets.

announcement that the kind of minerals suggested by the Opportunity spectrometers suggests an acid groundwater environment,

> one that is suitable for some kinds of organisms. The other thing is that with minerals precipitating out of a solution, those minerals as they precipitate can trap chemicals and organic materials and whatever else might be in the water at the time that the precipitation takes place, providing a very good long-term preservation mechanism. So if we can get our hands on some of these rocks and get them back to Earth and a laboratory, I think they would have a very interesting story to tell.

Three weeks after the announcement that Opportunity had found minerals indicating a watery past, JPL briefed reporters with an even more important discovery: some of the rocks seen at Opportunity's crater (by now named Eagle) were almost certainly sediments formed at the bottom of a shallow salty sea or lake. The rover was, in effect, trundling along an ancient beach. 'We think Opportunity is parked on what was once the shoreline of a salty sea on Mars', said Steve Squyres.

After taking hundreds of close-up photographs, Opportunity had discovered fine-scale bedding patterns in the fine-grained

The 2004 Mars rovers were equipped with drills to bore into the rocks they found. Previous landers were not equipped to study anything beneath the rock's outer 'skin', which may have been altered by chemical weathering or exposure to sunlight. These holes were drilled by Opportunity into the 2-m (7-ft) wide formation named El Capitan, and the subsequent chemical analysis of the exposed, fresh rock gave clues to this part of Mars's watery past.

rocks. Unlike the larger-scale cross-bedding seen earlier, these structures looked unambiguously aquatic in origin. The sediment particles, the size of sand grains, are shaped into ripples by water at least 50 mm (2 inches) deep, according to the JPL team. That does not sound much, but it is a minimum depth. The water could have been several meters deep, or more.

These fine bedding structures in rocks examined by Opportunity are further evidence that this part of Mars, a least, had a watery past

WHITE MARS: DEAD IN THE WATER?

'I have always thought that the final answer will be a compromise'

Nick Hoffman

While the newspaper headlines screamed babbling brooks, lakes, and life, the critics of warm, wet Mars urged that the Opportunity discovery be kept in perspective. According to Nick Hoffman, whose White Mars hypothesis posits that the planet has been essentially very dry and very cold for the whole of its existence, the discovery of these watery rocks rules out the 'hardest' White Mars position. At least some of the planet must have been warm and wet, at least in its early history. 'Since the Meridiani deposits are semi-regional, and we see similar "white rocks" (actually pale beige) in many other areas of Mars, it significantly downgrades my theories of a global CO_2-rich Mars', Hoffman says:

Furthermore, it was flowing – at around 100–200 mm (4–8 inches) a second. According to John Grotzinger, a rover science-team member, the environment at the time the rocks were forming could have been a salt flat, or *playa*, sometimes covered by shallow water and sometimes dry. Such environments on Earth, either at the edge of oceans or in desert basins, can have currents of water that produce the type of ripples seen in these Mars rocks. Jim Garvin believed that the follow-the-water strategy was paying off:

Many features on the surface of Mars that orbiting spacecraft have revealed to us in the past three decades look like signs of liquid water, but we have never before had this definitive class of evidence from the Martian rocks themselves. We planned the Mars Exploration Rover project to look for evidence like this, and it is succeeding better than we had any right to hope. Someday we must collect these rocks and bring them back to terrestrial laboratories to read their records for clues to the biological potential of Mars.

On the other hand, this is not yet 'proof' of lakes and oceans on Mars. One thing we do know about Meridiani is that it is not an enclosed basin – the hematite-rich deposits formed on a regional slope. So we can't have had a lake here . . .

As a result, I favor a 'Sabkha'-style environment for the formation of the initial sulfate deposits.' Sabkhas are low-lying desert flats where salts actively precipitate from groundwater, but are not the site of frequent lakes. The water table reaches the surface and evaporates, and minerals accumulate in layers. Nodules and tabular crystals are common in this environment. For Mars, I envisage groundwater seeping to the surface across the whole Meridiani region, and forming a smooth layered deposit that drapes everything – craters, hills, valleys etc, infilling

(*Opposite*) This rock, named Last Chance and seen here in a reconstructed three-dimensional view, contains structures that suggest it was laid down in water. Wiggly structures have been seen which have been interpreted as fossilized ripples, formed by flowing water.

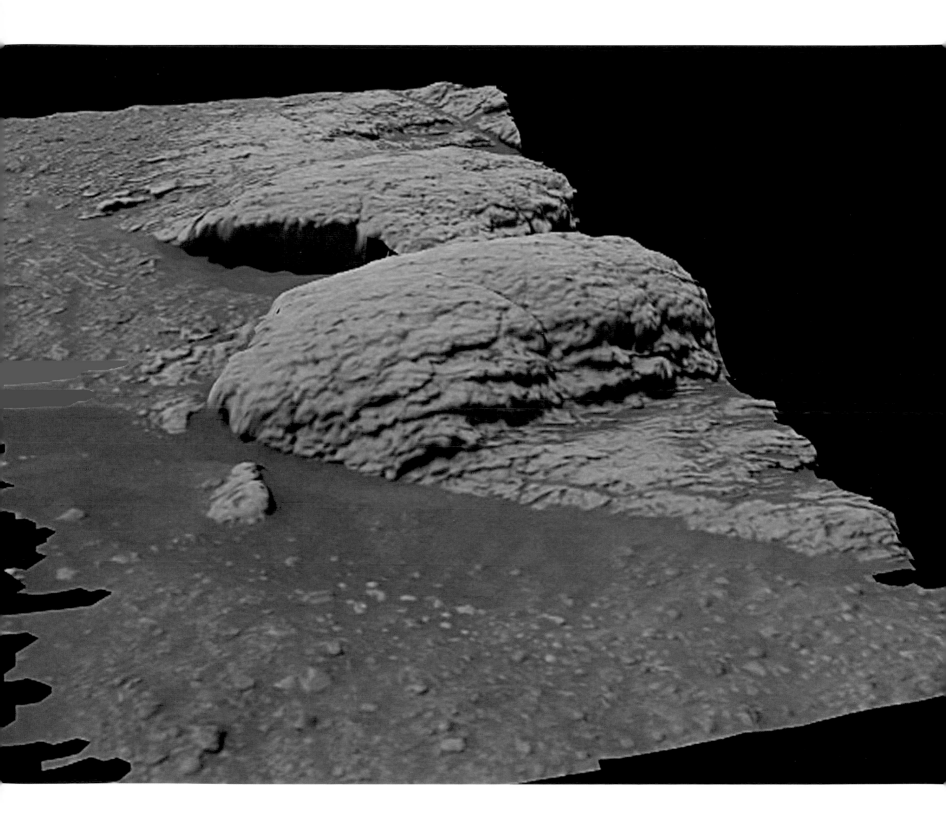

This magnificently detailed image taken by the panoramic camera on the Mars Exploration Rover named Opportunity shows the impact crater known as Endurance, which was reached after a drive of several hundred meters from Eagle crater, in which it landed. When the picture was taken, Opportunity was perched just 40 cm (15 inches) from the crater's edge. Endurance is roughly 130 meters (430 ft) across and about 18 meters (60 ft) deep.

the lows and smoothing the terrain. Any new craters would be rapidly filled-in. Of course, all this took place long ago and that groundwater is no longer active. Nonetheless, it Makes Meridiani a very interesting site, since the thickness of layered 'white' rock we see could have taken millions of years to accumulate.

Lakes several meters deep probably did not happen, says Hoffman; he believes any standing water at Meridiani was a few centimeters in depth at most.

White Mars lives on, but Hoffman admits that the evidence now points to a very mixed bag of processes operating on early Mars – and today. Meridiani is evidence of water, but the jury is still very much out on whether this water was in the form of a large lake or shallow sea, or whether it consisted of saturated, wet rocks – the water liquefied by a localized, temporary event such as volcanic activity or a meteorite impact. As to the large-scale channels seen on Mars from orbit, Hoffman admits that the latest photographs point to H_2O rather than CO_2. However, he is adamant that only some sort of carbon dioxide mechanism can explain the gullies

– the small-scale recent phenomena that have been interpreted by so many as evidence that liquid water still plays a role on Mars today. He says:

> Some in the community are satisfied that the Opportunity findings blow me out of the water and aren't even talking to me. But I believe in never giving up too soon. I have always thought that the final answer will be a compromise.

What does the Meridiani discovery say about the chances of life on Mars? The presence of ancient water increases the likelihood that life once took hold on the Red Planet. Sadly, the American rovers were not equipped to search for the sort of biogenic chemicals that could detect ancient or extant microbes. It is possible that the 'signature for life' could be lurking in the very rocks that Opportunity is, at the time of writing, drilling away at and photographing (some of the nodular formation in the rocks do look superficially like fossils, which has led inevitably to more speculation). But this spacecraft has no way of knowing. The search for life will have to be conducted by another mission, by a lander or rover fitted with a mass spectrometer like that carried by the ill-fated Beagle 2. And the search for life – which has always been the driving force behind Martian exploration – is now about to step up a gear. If scientists do find life on Mars in the next twenty years or so, it will be the discovery of the century.

Chapter 12
Martians

'Is there life on Mars?'

David Bowie ('Life on Mars', 1973)

Finding alien life would be one of the great discoveries of all time, if not the greatest. Knowing that alien life existed – be it microbial or advanced – would, in a sense, be the ultimate step of the Copernican revolution. The sixteenth-century Polish astronomer demoted Earth from its position at the center of the solar system. Later astronomers went on to show that our solar system was not the center of the galaxy, and then that our galaxy was neither unique nor at the center of the universe. To demote us further, Darwin moved from the pinnacle of creation. Martians – or any other extra-terretrials – would demote the whole of life on Earth from center stage. That is part of what makes the idea of alien life is so alluring.

When it seeped into the educated consciousness that other planets are worlds in their own right, it quickly became assumed that there were men living on them. The notion of an inhabited Mars was not the brainchild of Percival Lowell, but it was Lowell who first put flesh on the bones of the alien. One hundred years ago, thanks to Lowell, the Martian question made the newspapers. Today, thanks to the exploits of NASA and ESA, it still does. Of course, the debate has changed. Then, we wondered about the possibility of intelligent life. If

Relatively few illustrations of Martians have shown the archetypal 'little green men'. Perhaps the most original and imaginative Martians to date were dreamt up by H.G. Wells, whose invaders were wholly alien (and hence terrifying). This creature, an illustration by sci-fi artist Frank Paul for the May 1939 issue of *Fantastic Adventures*, has large lungs to extract the maximum amount of oxygen from the thin Martian air, and huge ears to capture the faint sounds carried in the planet's rarified atmosphere.

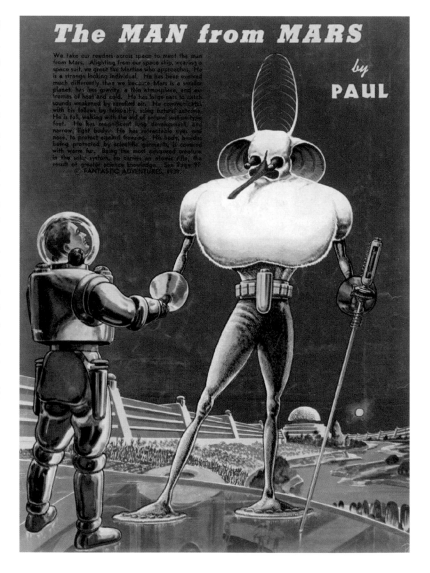

The MAN from MARS

by PAUL

there were Martians, what were they like? Socialist or capitalist? Militaristic or pacifist? Christian or heathen? The Martians we seek today are rather smaller and have no political or religious affiliations, but the essential question remains: we still do not know if there is life on Mars.

We do know that the other planets in our solar system are extremely hostile to life, or at least life as we know it. Mercury and Venus are both far too hot for any sort of biochemistry to function on their surfaces. The gas giants lack a solid surface, and although the existence of floating 'gasbag' organisms sailing serenely through the clouds of Jupiter has been postulated, it seems unlikely that the atmospheres of the giant planets contain the necessary chemical prerequisites for life. Pluto is surely too cold. That leaves the major moons of the gas giants – and Mars – as the only places where we might have a chance of finding life.

Recently, interest has focussed on Jupiter's satellite Europa (and, to a lesser extent, Callisto and Ganymede) as a possible abode of life. Europa, which is about the size of our Moon, has a surface consisting of almost pure water ice, deep-frozen to –180°C. But tidal forces generated by the interaction of Europa with its neighboring moons, and with Jupiter itself, knead its innards and create enough heat to melt the bottom of the ice crust, forming a sub-glacial, planet-wide ocean of brine perhaps 100 km (60 miles) deep. Europa was photographed in detail by the Galileo space probe in the late 1990s, and its ice crust shows evidence of cracking and flexing, supporting the idea that an ocean lurks beneath. Galileo found that Europa (and Ganymede, perhaps Callisto too) has a magnetic field, lending support to the idea that an ocean, or oceans, of brine may be swirling around under the surface. Astrobiologists now think that the Galilean satellites of Jupiter – minus Io, which is a volcanic hell – may be some the most likely places to find life in the solar system. Unfortunately, unless (as has been suggested) there is considerable recycling of material from the bottom of the ice crust to the surface, it may be extremely difficult to gain access to this life. If we need to drill through 20 km (12 miles) of granite-hard frozen water to get to Europa's hidden ocean, it may be a century or more before we have the required technology – and money – to mount such an expedition, even robotically.

Where else could life be hiding? Saturn has a huge moon, Titan, which is the only planetary satellite in the solar system to possess a substantial atmosphere – in fact, it has one and a half times the pressure of Earth's, and ten times the density. It is similar in composition to ours, in that the main constituent is nitrogen. But Titan's atmosphere is also rich in organic compounds, mostly short-chain hydrocarbons such as methane and ethane, which may condense and rain down onto its surface, pooling as oily hydrocarbon seas and lakes. According to Ralph Lorenz, of the University of Arizona, the surface of Titan could consist of bright 'continents' of water and other ices separated by large lakes, perhaps formed by impact craters and filled with liquid methane or somesuch. 'It would be a spectacular, but subdued landscape', he says. 'It would look something like Sweden from the air.'

'It would be a spectacular, but subdued landscape. It would look something like Sweden from the air'

Ralph Lorenz

Titan has all the ingredients for life – plentiful H_2O, a solid surface, lots of gloopy organic chemicals – but unfortunately it is rather cold. Surface temperatures probably do not exceed –180°C, and it is hard to see any exciting chemistry happening when the thermometer dips that low. This does not rule out life on Titan, however. Deep underground – many hundreds of kilometers beneath the ice-continents and methane lakes – there may be a bizarre parallel world consisting of another planetary ocean, this time made of liquid water. By the beginning of 2005, the Cassini probe is due to arrive at the Saturnian system and drop a European lander, named Huygens, into the clouds of Titan. In the couple of hours or so Huygens will have before its batteries run out, it will hopefully take some spectacular pictures of its descent and make some preliminary analyses of Titan's surface and atmosphere. There is a good chance that Huygens may even splash down in a hydrocarbon lake – it should float for a short time. But it won't find life – unless it is of the spectacular and obvious kind. So that just leaves Neptune's large moon Triton, which appears to have some interesting geological activity going on underground, and Mars.

In January 2005, the Huygens probe was due to arrive at Saturn's cloud-veiled moon Titan. It is thought that much of Titan's surface is covered with lakes or seas of dark, oily hydrocarbons, into which the probe may well splash down. Titan has the ingredients for life, but it is very cold.

COULD LIFE EXIST ON MARS?

If there is life on our planetary neighbor, it is probably not the sort of life we are familiar with. At present the surface of Mars would be lethal for all terrestrial life-forms, however hardy. The atmosphere is a near vacuum, and surface pressures nowhere exceed 10–12 millibars. The Martian air is almost pure carbon dioxide, and temperatures are sub-Antarctic over much of the planet. Any terrestrial microbe, even those adapted to live in the coldest niches of the Antarctic, would at Martian temperatures suffer fatal cell damage through the formation of ice crystals.

Apart from the life question, nothing is more controversial than the debate over the existence liquid water on the Martian surface, but everyone agrees that Mars is generally far drier than the driest deserts on Earth. Liquid water is not stable at Martian pressures and temperatures, and most astrobiologists find it hard to postulate a metabolism not based on water. Mars is also bathed in ionizing ultraviolet radiation in quantities that would cause severe damage to complex organic compounds such as DNA. The surface of Mars also appears to be rich in bleach-like chemicals called superoxidants, which are very damaging to life.

So the fate of an Earthly organism would be a rapid death on the surface of Mars – it would be deep-frozen, desiccated, and fried simultaneously – but this does not preclude *Martian* life. Organisms that had originated on Mars would naturally have evolved to exist in the prevailing environment; and even if that environment changed dramatically (as it seems that it may have done from time to time), the engine of natural selection may have been able to keep up, just as Earthly organisms have managed to colonize virtually every niche imaginable on Earth.

And there is far more to Mars than the surface. Just 10 m (30 ft) underground, conditions are far more clement. Liquid water could exist, the soil may retain some heat, and there is complete protection both from ionizing solar radiation and from the corrosive chemicals on the surface. And the possibility that Mars is still geologically active (plus the fact that the soil appears to be full of ice) means that life could eke out an existence clustered around the Martian equivalent of hot springs, if there are any. Life *could* exist on Mars, but so far there is only highly debatable evidence that it does. And the

fact that some terrestrial organisms may be able to survive in some protected Martian niches is not proof that they or their ancestors could have evolved there. After all, humans could quite certainly live on Mars supported by the appropriate technology, but we could not have evolved there.

From a distance, Mars does not look like a living planet. As James Lovelock, father of the Gaia Hypothesis, points out, the existence of a Martian biosphere would have had profound effects on its atmosphere and lithosphere. Planets with life are likely to have chemically very unstable atmospheres, which contain large quantities of reactive chemicals such as oxygen and methane (in fact, small quantities of the latter gas have recently been discovered in Mars's atmosphere – as discussed later in this chapter). Nevertheless, some have pointed out that there may be very circumstantial evidence of past life on Mars, at least. The redness of this planet is down to oxidation. Could an early flourishing of life have produced enough methane to turn the rocks red?

It seems very odd now that the notion of higher life-forms on another world was such a casual matter during the Lowellian age and before. In the seventeenth century, Christiaan Huygens, in his book *Cosmotheros* (published posthumously in 1698), wrote that all the planets in the solar system were inhabited. And in 1784, William Herschel was the first to describe the melting of the Martian polar caps in springtime, suggesting that they may be made of ice or snow. He concluded that Mars has a thin atmosphere but a basically Earthlike environment, sowing the seeds for two centuries of speculation on the existence – and nature – of Martians.

'This opened up a real Martian can of worms'

Monica Grady

THE METEORITE FROM MARS

On 27 December 1984, Roberta Score, a member of a scientific team scouring the icefields of the Allan Hills region of Antarctica for meteorites, stopped to admire a particularly fine view of some ice pillars, glinting in the midsummer sun. Scientists come to Antarctica to look for meteorites because it is probably the easiest place in the world to spot them.

Any rock lying on the ice must have come from the sky, and their darkness makes them stand out clearly. Deserts are another favored location for meteorite-hunters. Score spotted a green rock lying on the ice. She picked it up and sealed it in a bag. She later decided that it was probably the most interesting of the 1984 meteorite collection, and assigned it the number one, and thus it became known as ALH (from Allan Hills) 84001. The meteorite was then sent off to the NASA labs in Houston, where it was put in a cupboard and, for the time being, quietly forgotten. The story of ALH 84001 is instructive as it illustrates the passion of those who wish to see Mars live.

Eight years after its discovery, ALH 84001 came to light when it was examined by members of the Antarctic Meteorite Working Group, an international team of scientists who investigate these rare and pristine pieces of the solar system. When the scientists started to study it, they saw straight away that it was unusual. It was classified as an orthopyroxenite, basically an iron magnesium silicate igneous rock. It was first assumed that the rock originated from the asteroid Vesta, as meteorites with similar composition to ALH 84001 had previously been identified as having come from there. But the Antarctic rock had some extra features of interest, namely some strange orange patches of carbonate minerals. Again, this was not totally unexpected; it was known that carbonates could leach into igneous rocks after exposure to solute-laden waters. Perhaps ALH 84001 had acquired the orange minerals during its long sojourn in Antarctica.

Then, in 1993, David Mittlefehldt, one of the JSC NASA team, announced that this rock was from Mars. Analysis of gas trapped in pockets in the meteorite showed that the ratios of certain gases present tallied perfectly with those sniffed in the Martian atmosphere by the Viking landers. The evidence that ALH 84001 is a piece of Mars is certain and wholly uncontroversial. But that is the only context in which you can describe this piece of stone as 'uncontroversial'.

A consortium was formed to study the meteorite, which had quickly become one of the most valuable specimens in NASA's rock collection (at that time only ten or so Martian meteorites were known to science). Three of the team consisted of British meteorite experts Monica Grady and her husband Ian Wright, and Colin Pillinger, the man who went on to achieve the wrong sort of fame as the father of the ill-fated Beagle 2 mission. Presenting their results at a conference in Prague in 1994, the British team suggested that ALH 84001 had had a very interesting history. 'There was no suggestion of life,' Grady says, 'but this meteorite was very unusual, very interesting.' Shortly after the conference, Grady published a paper in *Nature* suggesting that the carbonates in the Allan Hills rock had been laid down in warm water.

Skip forward two years to the Meteoritical Society's annual meeting, held that year in Berlin. According to the grapevine, a big discovery had been made about the meteorite from Mars. 'People were talking in corners, trying to find out what was going on', Grady recalls. 'The rumors said that the Americans had found fossils in ALH 84001.'

In fact, ALH 84001 was not the first Martian meteorite to excite the interest of astrobiologists. Another rock, designated EETA 7901, had already been found in the early 1990s to contain organic chemicals associated with possible Martian carbonates, but it was the news that NASA might be about to announce not mere carbonates but actual fossils in ALH 84001 that really got the world excited. Someone leaked the news that this was precisely what NASA was claiming – to the BBC (perhaps it was thought more politic to leak the story to a foreign news source than risk federal wrath by leaking to the US media).

On 7 August 1996 the BBC announced that Martian fossils had been discovered in ALH 84001. A paper entitled 'Search for past life on Mars: Possible relic biogenic activity in Martian meteorite ALH 84001' was to be published in *Science* the following week, but to the great annoyance of its publishers the news was already out. That week, Everett Gibson and David McKay, the lead authors of the *Science* paper, announced their findings, and after a briefing by NASA administrator Dan Goldin, President Bill Clinton announced the historic discovery of life (albeit dead life) on Mars. Suddenly, a planet that had been written off as dead ever since Mariner 4 had showed that much of it looked like the Moon found itself back in vogue. 'This opened up a real Martian can of worms', Grady says.

Ever since, the debate has raged over what exactly McKay and his team had found. Every year or so, it seems, a story emerges either confirming or disproving the conclusion that there are indeed alien fossils in the rock. The 'exciting' conclusion is based on several findings. The first is that the rock

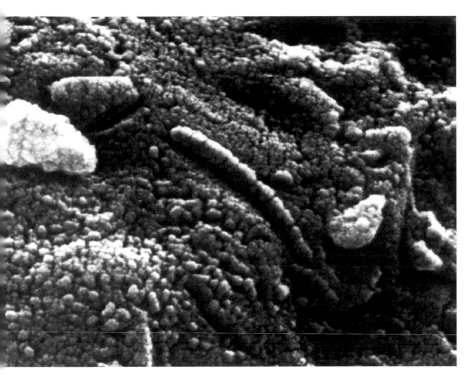

The Martian meteorite ALH 84001 contains mysterious tubular structures that some scientists believe to be fossil bacteria. If they are, then they are an order of magnitude smaller than any terrestrial equivalent.

is indisputably from Mars. It was probably blasted off the Martian surface several million years ago by the impact of a much larger meteorite. Planetary scientists now know that several tonnes of Mars rock come winging their way across space every year towards Earth. Most of these fragments are burnt up in our atmosphere, but a few large chunks survive their fiery plunge through the air to be spotted by beady-eyed meteorite hunters. The second plank of evidence is that the carbonates formed on Mars, not on Earth. This again is not that controversial. The isotopes of carbon found in the deposits in ALH 84001 are consistent with those found in the Martian atmosphere. Thirdly, it was assumed that the carbonates were formed under water, but that is only one explanation. They could have been rammed into the rock during the impact that expelled the meteorite, or they could have formed in wet – but cold – conditions. A fourth finding is that ALH 84001 contains compounds called polycyclic aromatic hydrocarbons

(PAHs), which can be created by bacteria, though many scientists believe these are simply contamination from Antarctica. But the most 'conclusive' evidence for the mini-Martians consists of tiny grains of magnetite – a magnetic oxide of iron – in the meteorite. Magnetite is a mineral that commonly forms as a by-product of bacterial action. Microbes often leave chains of magnetite crystals. The NASA team said they had identified several such chains in ALH 84001. 'But they had *not* provided good evidence for this', says Grady. 'This is where the argument is raging the hottest.'

Now, most people believe that the magnetite, although incontrovertibly present, was not formed by life but was deposited by purely inorganic processes. The argument is getting quite vitriolic, but the 'non-life' brigade is winning the argument. In the end, whether you believe ALH 84001 contains the remains of Martian bacteria, which perhaps once thrived on an ancient seafloor before the planet's climate went to pot, is something of a matter of faith. 'If you were to put the question to fifty astrobiologists,' Monica Grady says, 'forty-nine would say these structures are inorganic.'

What really caught the public imagination about ALH 84001 were not arcane arguments about biogenic magnetite crystals and isotopic carbon, but the pictures of tiny wormlike structures seen in the meteorite (these were used as the peg for a rather entertaining episode of the *X-Files* in which hostile Martian microbes emerge from ALH-type meteorites, with unhappy results). These look for all the world like the sort of segmented tubular bacteria familiar on Earth. The problem is, they are at least one, possibly two orders of magnitude smaller than any known living cells – the ALH 84001 'micro-fossils' are about a couple of hundred nanometres long (about 1/100,000 of an inch). This is far too tiny, say most biologists: there simply is not the room for all the machinery of life – for reproduction, respiration, and so on – to be packed into such tiny chambers. Never mind, said those who maintained that they *were* fossils. Perhaps these structures had been squashed, dried out, shriveled in some way.

When, in the late 1990s, scientists started to discover minute structures in cores from deep-sea drilling, the Mars meteorite community pricked up its collective ears. Phillipa Uwins, a British geologist working at Queensland University, Australia, caused something of a stir back in 1999 when she claimed that rock samples from several hundred meters under the Indian

Ocean floor contained virus-sized structures that appeared to be alive. These 'nanobes', as they have become known, are as small as viruses yet, under the microscope, looked like tiny bacteria – and, unlike viruses, appeared to be capable of independent existence and multiplication. Indeed, she spoke of the things multiplying forth all over her lab equipment. Since then, doubt has been cast in some quarters whether these nanobes really are alive, and conclusive proof in the form of DNA testing has still not been forthcoming.

Nevertheless, if Uwins really had discovered a hitherto unknown form of life – maybe a whole new kingdom – then the probability that the little wormlike things in ALH 84001 might once also have been alive increases dramatically. Any evidence that terrestrial life can deviate massively from the norm is warmly welcomed by the astrobiologists. Olavi Jajander, a Finnish scientist, also thinks he has found nanobes. Perhaps the most welcome news for the neo-Lowellians has been the discovery that Earthly life can exist in all sorts of environments once considered incontrovertibly too hostile for biology. To find 'life Jim, but not as we know it', you no longer need the fire up the warp drive and put several light-years between yourself and Earth, you just need to visit an undersea vent or a hot spring, where bacteria have been found thriving at temperatures in excess of 130°C and pressures of thousands of atmospheres. It may in fact be the case that the greater part of Earth's biosphere is located not on the surface or in the oceans, but deep underground, in the tiny pore spaces in the rocks. You have to go quite a long way down before geothermal heat kicks in to cook anything that could reasonably be expected to be alive, and if the top kilometer or so of the Earth's crust were saturated with tiny bacteria, the total biomass of this planet could be an order of magnitude greater than was thought. This raises a big question: if the most plentiful terrestrial life-form is a rock-dwelling nanobe, is this way of life a 'recent' adaptation from ancestry that evolved and dwelled upon the surface, or are we all the descendents of nano-sized extremophiles – organisms that can survive in environments hitherto thought to be hostile to life, such as superheated or highly acidic water? And if the Earthly abode of life is much larger than was once thought – from several kilometers beneath the surface to almost the top of the atmosphere, then this increases the likelihood of life existing also on Mars.

'*who is to say that the rocks weren't staring right back?*'

Henry Gee

In March 2000, a paper in the *Proceedings of the National Academy of Sciences* by Caltech's Benjamin Weiss suggested that buried Martian life-forms could live off hydrogen and carbon monoxide, produced by ultraviolet radiation acting on water and carbon dioxide on the surface and percolating down through the regolith. This life should in principle be detectable, they argued, not directly (which would involve digging down through tens or even hundreds of meters of rock) but by detecting the subtle imprint of this subterranean ecosystem on the Martian atmosphere. If the quantities of carbon monoxide and hydrogen detected in the atmosphere differ even slightly from what we would expected for a wholly inorganic environment, then this would be circumstantial evidence of life.

The story of ALH 84001 tells us one very useful thing about Martian life – even if the 'microbes' within it turn out to be microscopic red herrings. And that is that Martians are probably going to be very difficult to spot even when they are right under our noses. With structures – particularly mineralized fossil structures – that are just a few tens or hundreds of nanometres across, the difference between the organic world and the inorganic is a fine one indeed. At this scale, a great many things can be mistaken for life – crystals, gas bubbles, inorganic nodules, and concretions, to name a few. It is only when life starts moving around and waving at us that we can say for sure what it is. Even on Earth, the line between the living world and the non-living is not as clear-cut as we sometimes think. What are Phillipa Uwins' nano-things? Exotic, claylike minerals, or tiny bacteria? What, for that matter, are viruses? Dead or alive? Would a Martian virus be alive? How would be recognize a virus fossil? Throughout the life on Mars debate the Lowellian paradigm that the Martians would, in essence, be recognizable to us as part of the universal kingdom of the living has not been seriously challenged. As Henry Gee, a journalist on *Nature*, wrote recently,

there is no need for life anywhere else in the Universe to look like terrestrial life in any way at all. The Earth is the

Earth, and Mars is Mars. When the Mars rover [Sojourner] came eye to eye with rocks on its brief scuttle over the surface of Mars, who is to say that the rocks weren't staring right back?'

ARE WE THE MARTIANS?

When H.G. Wells imagined the Martian invasion of England, he wrote of a technologically advanced civilization which constructed spacecraft to get across the void of interplanetary space. But according to some scientists, accomplishing such an invasion might be a lot easier – and you would not need any sort of technology at all to do it. And the key to this evolution lies in meteorites such as ALH 84001.

An entertaining and quite plausible theory is that, given the amount of material of Martian origin known to strike Earth every year, it is at least possible that had life originated on Mars then some of it could have been carried, intact, across interplanetary space and found a new home here. Four billion years ago both Mars and Earth were the subject of heavy cosmic bombardment from the planetary leftovers that swarmed through the early solar system. If life arose on Mars at around this time, chunks of rock impregnated with Martian microbes or spores could have been blasted off the surface of their home planet and made their way to Earth, where they took hold. If this was the case, then the entire biota of our planet would have originated from an alien genesis – and we would all be Martians. Instead, of course, the reverse could have happened. Earth was also struck time and time again during this distant epoch, and early earthlings could have made their way to Mars.

But according to Paul Davies, an astrobiologist at Macquarie University, Sydney, the chances are that if an interplanetary seeding took place, then it was Mars that seeded the Earth not the other way round. 'Up to now,' he says, 'everyone has assumed that because there is life on Earth it must have started on Earth. But there is no reason to believe that this is so.' This is a bizarre theory which many will dismiss out of hand. But since both Davies and another scientist, Jay Melosh of Arizona University, came up with it completely independently in 1992, no one has yet managed to knock it down. As Christopher McKay of NASA's Ames Research Center once memorably put it, Mars and Earth have been 'swapping spit' for billions of years.

Davies' argument rests on the fact that fossil (chemical) evidence for primordial life has been found in Greenlandic rocks of extreme antiquity – dating back some 3.8 billion years. The trouble is that if these rocks really do bear the chemical signature of life (and the evidence is, according to many, as shaky as for Martian microbes in ALH 84001), then where did they come from? If they are truly fossils, they must be of organisms already several stages removed from the primordial soup (or hot vent, or wherever). That pushes the date for a terrestrial genesis back beyond the 4-billion-year mark. And that is problematic. Earth at that time, the appropriately named Hadean period, was undergoing a ferociously heavy bombardment. Asteroids 100 km (60 miles) across were slamming into its surface at 10-million-year intervals. Just as life got started, if at all, it would have been wiped out during the next strike. Yet here, dating from just a couple of hundred million years later, we apparently have life which is capable of leaving a mineralogical signature.

The young Earth was a hellish place but, Davies has suggested, the young Mars may have been far more hospitable. Being smaller and lighter, it would have been a smaller target for the asteroids to hit, and with less gravity to attract them. With so much water on its surface, Earth would have been converted into an autoclave after each major impact, the planet-wide ocean vaporizing into an atmosphere of superheated steam. And because Mars is smaller than Earth, it generates far less geothermal heat. On Earth, primitive bacteria could in theory have sat out the bombardment by migrating underground. But the options for doing this were limited. Even a few kilometers below the surface the young Earth, the rocks would have been too hot for any conceivable type of life to exist. But on Mars, the subsurface rocks would always have been cooler, allowing Martian bacteria to shelter far more easily from Mars's own (not inconsiderable) bombardment. And up on the surface – if the warm, wet theory is right – the young Mars would have been a far more hospitable place than it is today, with a thicker, milder atmosphere and large bodies of standing water. Just the right pace, in other words, to cook up a primordial soup. 'Mars was simply a better place for life to get started back then', says Davies. In fact, the basaltic rock which constitutes ALH 84001 is actually older than the mother-of-all-impacts that nearly destroyed Earth and led to the creation of the Moon. Some scientists have even suggested that the three great branches of the kingdom of life – the prokarya, archaea, and eukarya – may represent three distinct waves of Martian invasions. And if Phillipa Uwins' nanobes are confirmed as being alive, they may be a fourth.

It's a nice theory – and especially attractive because it is eminently testable, provided we can find some sort of extant Martian life. A DNA test would soon tell us whether our life and Martian life were brothers under the same Sun. However, there are a lot of ifs and buts. Making the early Mars warm enough and wet enough to be a plausible garden of Eden is, as Nick Hoffman and others have pointed out, extremely difficult. There is the young dim Sun problem, and also the problem of where would Mars get all that atmosphere from. It is far from certain that life on Earth did start as early as 3.8 billion years ago. Those Greenland 'fossils' could be inorganic artifacts. Finding traces of dead creatures billions of years old is hard enough; if they were creatures on the nanometre scale, it becomes exponentially more difficult.

Imagining a bacterial spore surviving the rigors of a million-year spaceflight is also pretty hard. But claims have been made about viable Earthly bacterial cysts that have been found in salt crystals that are a quarter of a billion years old, and if the claims are true, then microbial life turns out to be incredibly robust. It is only big complicated creatures like us that need to be too worried about the rigors of interplanetary vacuum and stellar radiation. If Davies is right, our two planets would be united in true kinship. Although whether this would lead us to teat our original home with any more respect than we have treated our adopted one is a matter of doubt.

THE VIKING CONTROVERSY

To date, the only successful probes to carry instruments capable of detecting life in situ were the JPL Viking landers, which were put down on opposite sides of Mars in the summer of 1976. They were each equipped with a battery of automated experiments designed to test for the presence of living microbes in the Martian soil. On 28 July, a robotic arm reached down from Viking 1 and transferred some soil to the onboard biology laboratory. In the so-called Labeled Release Experiment, samples of Martian soil were incubated in an atmosphere containing carbon dioxide – the carbon component being the radioactive isotope carbon-14 (other

'If a dog had shit on the ground one meter from a Viking lander, it would never have detected it'

Fraser Fanale

forms of carbon, which naturally occur on Earth and Mars, are not radioactive). It was hoped that if life was present, it would absorb the carbon-14 from the air. Then the air would be evacuated from the experiment chamber, and the soil would be cooked. This would release the carbon-14, which could then be detected through its radioactivity. In addition, water was added to Martian soil and the gases given off were analyzed. The soil was then cooked and analyzed by a mass spectrometer to try and detect organic chemicals.

The results were mixed. While several of the automated experiments – notably the Labeled-Release Experiment – tested positive, the mass spectrometer found no evidence of organic material. If any was present, it must have amounted to less than one-hundredth of the quantity of life-stuff that you would find on the most sterile desert on Earth. Furthermore, when water was added to Martian soil, large quantities of oxygen were given off. This could be interpreted as the action of microbes – or it could be interpreted as the action of superoxide chemicals in the soil, formed by aeons of exposure to ultra-violet light. Occam's razor was duly wielded, and it was concluded that Mars was dead.

But not everyone is convinced that Viking came up with a negative result. According to one report, at a 1997 NASA meeting on Martian life, geochemist Fraser Fanale opined 'If a dog had shit on the ground one meter from a Viking lander, it would never have detected it.' This is a comment on the experimental design, not on the presence of Martian life, but some scientists go further. Gil Levin, a former mission scientist on Viking and the principal investigator on the Labeled Release Experiment, says that the spacecraft almost certainly *did* discover life. His argument is that the organic analysis instrument – the results from which were official clincher that Viking did not find bugs – was too insensitive, requiring millions of microorganisms to show a positive reading, whereas the Labeled Release Experiment – which *did* give a positive result – was capable of detecting activity from as few as fifty microbes. And to cap it all, when the sample was cooked and retested,

If life – or fossil life – is ever discovered on Mars, it may well be by a human rather than a robot. Humans are very good at seeing patterns and simply make better, more intuitive, and more resourceful explorers than any machine could manage in the near future.

all signs of 'life' had disappeared, suggesting that the machine was simply not recording the result of inorganic chemistry. A statement on Levin's website reads:

> In my 1986 talk at the National Academy of Sciences' tenth anniversary of Viking I presented all the available evidence bearing on the LR results. I concluded my analysis with the statement that 'more likely than not, the LR discovered life on Mars.' Following the publication of news about the first Martian meteorite (ALH 84001) evidence, I told a calling reporter that, were the analyses confirmed, I would change my conclusion to 'most likely.' When the second meteorite (EETA 79001) results were announced, upon inquiry by another reporter, I said (again, presuming the report valid) 'almost certainly'.

Levin goes on to claim that Viking's cameras saw greenish patches on some Martian rocks, and he argues persuasively that at both Viking sites pressures and temperatures were such that, for quite long periods of time, liquid water could have existed at these sites.

WANTED, MARTIANS: DEAD OR ALIVE

Viking's results were ambiguous, that much is clear. But Viking did answer some vital questions about what sort of experiments are needed to detect life on Mars. As there is not (probably) going to be any macroscopic evidence (green patches excepted), any spacecraft searching for life is going to have to perform some pretty ingenious experiments in order to detect microbial activity. Searching for evidence of past life will be even harder. The first spacecraft to be sent to Mars after the Vikings specifically designed to look for life was Beagle 2.

Sadly, Beagle failed. What about the Mars Exploration Rovers? When they were launched, some commentators enthusiastically sold them as 'NASA's new search for life on Mars', whereas in fact they were equipped only to search for signs that the Martian surface may once have been the sort of environment where life could have existed. The presence of minerals such as jarosite has been taken as fairly conclusive proof that this part of Mars at least was once wet – or at least wettish. Even a damp salt pan is a world away from the sort of hellish desiccation suggested by Viking. Jim Garvin's statement that this was exactly the sort of result anyone wanting evidence for past life was hoping for seems to reflect current NASA thinking. We haven't found life – dead or alive – on Mars yet, but the odds that someday we will are improving all the time.

STOP PRESS: LIFE FOUND?

In March 2004, intriguing reports started to emerge about the presence of methane in the Martian atmosphere. Methane is considered to be one of the strongest possible indicators of Martian life – it is a gas known to be produced biogenically on Earth, and it is highly unstable in the Martian atmosphere, being degraded by the action of sunlight within a few hundred years. If substantial quantities of methane were to be found in the Martian air, it could mean only two things: either there is extant life, probably living under the surface, churning the stuff out, or there is a supply of methane in some still-active volcanic vent. And an active volcanic vent may well support life, so one suggests the other.

Preliminary data from the Planetary Fourier Spectrometer aboard Mars Express has found spectral emissions of methane in the Martian atmosphere. Vittorio Formisano, the man in charge of the instrument, told *Science*, 'We have seen methane on Mars.' The team, led by Vladimir Krasnopolsky of the Catholic University of America, added in a short paper that what had been found was:

A very little amount, but the result is clear. Methane on Mars has been detected from the Earth as well. A team using the Fourier Transform Spectrometer at the Canada–France–Hawaii Telescope has also spotted the gas. Methanogenesis by living subterranean organisms is the most likely explanation for this discovery.

CONTAMINATION

The confirmation of Martian life would have profound consequences – for the future of space exploration, for the future of science, and for the way we think about ourselves. Because if we find life on Mars, and manage to show that this life is not simply the result of cross-contamination (either way) with the Earth, it will imply that a second genesis occurred on the first planet we went looking for it. At the moment we know of only one planet with life. All our assumptions about what life is and what it does to a planet are based on this one example. But if there is – or was – life on Mars, we will instantly double our biology database. But most importantly, if there is life on Mars, then there is almost certainly life everywhere. We now know of more than a hundred stars with orbiting planets outside our solar system. It seems probable that planets are the norm rather than the exception. If two, quite dissimilar planets in our solar system alone independently evolved life, then this would suggest that life is not a particularly hard thing to get started. Our galaxy – and the universe – must probably be teeming with the stuff. The importance of Martian microbes would be not just in the organisms themselves (although this would be the most exiting news for biology since the discovery of the double helix), but what they would say about the ubiquity of life in the universe as a whole. All of a sudden, we would be Not Alone, and we would not have to wait for a radio signal from more advanced life forms to conclude that the probability of there being intelligent life out there would also have gone up enormously. The sheer magnitude of such a discovery explains why NASA is throwing so much effort into exploring Mars right now.

Paradoxically, if one of the robotic probes launched in the next twenty years or so does discover Martian life, it could also be a huge setback for Martian exploration. Astronauts are messy, however well insulated they are in their habitation modules and spacesuits. It would be a terrible tragedy to discover a whole new ecosystem, only to wipe it out with some carelessly discarded rubbish, swarming with Earthly nasties. Then there are those who would be worried about the possibility of life from Mars causing trouble here on Earth. Expect protests from the usual quarters (the Green movement) when the first sample-and-return missions are given the green light towards the end of the 2010s. The fact that Mars and Earth have been, as we have seen, 'swapping spit' for aeons without any obvious consequences will not deter those who will call for such missions to be banned.

In May 2002, *Nature* carried news of a report commissioned by the US National Academy of Sciences which highlighted the environmental risks faced by future astronauts on Mars. If the soil really is full of peroxides, then it could be extremely toxic and corrosive to exposed flesh. There is no way that large quantities of Martian dust would not make their way into the explorers' living area. The Apollo astronauts complained that the inside of their capsules were soon black with dust, which got into every nook and cranny of their bodies. Dust that is highly corrosive would be far worse. And to guard against 'back contamination' – bringing a Martian plague to Earth – the report recommended that humans should not be sent to any site not previously shown to be free of organic material, although it recognized 'that this may be in conflict with one of the primary goals of the exploration of Mars: to find extraterrestrial life' (*Nature* 417, p110, May 2002).

But what if there is no life on Mars, and never has been? Would this be such a disappointing discovery? The chances are that science will never be able to come to this conclusion. (In reality, the only two possible results are *yes* or *maybe*. Proving a negative on such a global scale will not be easy.) Although it would not have the philosophical connotations of a positive result – no life on Mars would not mean that we are not alone, it would just not be more-or-less conclusive proof that we were, which is a different matter entirely – no life on Mars would in many ways be just as interesting. Here we would have a planet with all the essential ingredients for life in place. There is good evidence that in their youth, Mars and Earth were quite similar. Mars may once have been warmer and wetter than it is today. All the chemicals needed for life are there in abundance – Mars is not the Moon. So if Mars has always been sterile, it would say some quite profound things about life on Earth, and some quite profound things about how rare life might be and how difficult it is to get started. A dead Mars would be a very clear sign that our planet is even more special than we thought. But even if Mars is dead today, it may not remain so for ever.

Chapter 13
EARTH ON MARS

'The Martians were there – in the canal – reflected in the water. Timothy and Michael and Robert and Mom and Dad'

Ray Bradbury (*The Martian Chronicles*, 1951)

Percival Lowell saw in Mars a world in essence much like the Earth. It had air you could breathe, clouds and winds, and the unmistakable presence of intelligent life. Mars is dead, but what if we could make it live again? What if we could create Earth on Mars?

Kim Stanley Robinson's *Red Mars* trilogy has as its central theme the colonization and eventual transformation of the planet Mars, from a cold, hostile, radiation-bathed desert to a far more clement place, with a breathable atmosphere of reasonable thickness, clouds and rain and rivers. This is achieved by a process known as terraforming – making an alien world like the Earth – and to many Mars enthusiasts the creation of a second home on the fourth planet is both a desirable and an achievable goal. Mars would then become the new frontier, a Wild West in space without the moral and military inconvenience of an indigenous population to deal with. The new Mars would become our second home, the first step outwards to the eventual conquest not just of the solar system, but of perhaps the galaxy as well. As well as the manifest destiny argument, which is essentially emotional, the creation of a second human habitat on Mars is often put forward on purely logical grounds: the chances are, advocates say, that sooner or later Earth will once again be struck by a large piece of space debris, as apparently occurred 65 million years ago off the coast of what is now Mexico. The consequences for the dinosaurs (if the Mexican impact really was responsible for their demise, something which is now being called into serious question) were dramatic. The consequences for us from a similar strike would be equally grim. Our species might survive, but our civilization would

Some research suggests that the Martian day – about 40 minutes longer than ours – and the otherworldly color of its skies (seen here from the Spirit rover's landing site) may be enough to make living on the planet for any length of time very difficult. We shall see. The human body, from the time of polar exploration, has proved itself to be remarkably resilient to all sorts of stresses and strains. Some scientists believed, in the early days of spaceflight, that more than a few days of weightlessness could have terrible effects.

surely perish. Humanity would be literally bombed back into the Pleistocene. We may be able to develop the technology to destroy or deflect such an asteroid, but as an insurance policy it would be extremely good if we were to create a second home as a backup. Just as it is always a good idea to back up valuable computer data on disks or memory sticks in case the original data is lost for some reason, creating a backup of our culture and knowledge base (not to mention our gene pool) on Mars would be a good idea. The chances of both planets being hit in any conceivable timescale are much smaller than for either one on its own.

Those are the reasons for terraforming. But how can it be achieved? A 1997 paper by Julian Hiscox, a British microbiologist, outlined how a process called ecopoiesis could be achieved on Mars. Ecopoiesis (the 'making of an ecosystem') differs from terraforming in that it involves not the creation of a replica Earth, but a limited transformation which would allow any extant Martian life (which we must assume is probably underground) to colonize, or recolonize, the surface. If it turns out that Mars is sterile even at depth (something we may never know), then the organisms needed to bring about a change to a warmer, wetter Mars would need to be imported.

Early ideas on terraforming Mars were often simplistic. One clever notion was to spread large quantities of soot or some other black dust over the polar caps so that the extra heat absorbed would hasten their sublimation, thickening the atmosphere in the process. The problem is that millions of tonnes of soot would be required – and how would we get it to Mars? It now turns out that much of the polar caps consist of water ice rather than dry ice, and sprinkling soot on ice at −100°C is not going to have much effect. In his book *Terraforming: Engineering Planetary Environments* (1989), Martyn Fogg suggested that vast nuclear explosions might be enough to melt the ice beneath the surface of Mars and in its polar caps, releasing an ocean. This would work, but it would be extremely expensive.

Cheaper, and probably easier, would be to construct huge orbiting mirrors to redirect and focus sunlight directly at the polar caps, leading to the release of CO_2. Carbon dioxide is a greenhouse gas, but not a terribly effective one compared with other substances such as chlorofluorocarbons (CFCs). Industrial plants constructed generating and pumping large quantities of CFCs into the Martian atmosphere may be able to warm the planet considerably in a matter of centuries or even decades. The beauty of terraforming is that it tends to be a process with a lot of positive feedback: warm Mars a bit, and the results will be such as to warm it further. There are many terrestrial organisms which could be candidates for Martian pioneers, such as hardy Antarctic microbes that can survive temperatures and pressures almost as low as those they would encounter in their new home. Tweak them with a bit of genetic modification, and they might even be even happier on Mars. To terraformers, plants are the ultimate planet-transforming machines. They eat carbon dioxide and produce oxygen, they stabilize the climate and extract and concentrate minerals from the soil. First microbes, then a few hardy lichens, then some squat shrubs. Finally, with factories churning out the CFCs that would have long been banned on Earth, Mars would have completed its transformation from red to blue to green.

Theoretically, all this could be done. The accomplishment of ecopoiesis would lead to what has been called Vitanova, the planet of new life (Terranova, the new Earth, would have to come later). It has been calculated that from the establishment of a primitive biosphere consisting of anaerobic microbes to the creation of an oxygen biosphere may take between 10,000 and 21,000 years. With the willpower, humans could build themselves a new home in roughly the same amount of time it took us to escape our African nursery and build the civilization we see today. And who can imagine the life future Martian colonists might lead? Cut off from Earth by millions of kilometers of space, they would doubtless evolve a whole new culture, new languages, and new philosophies. If we become Martians, then who knows what sort of planet we would choose to build? Maybe a Martian-Human civilization would live by the kind of high-minded, meritocratic ideas postulated by Lowell.

So much for the imagination. What about the reality? Terraforming Mars, even partially, may be technically feasible. It may even be desirable – the 'insurance policy' argument is a good one. No self-respecting technological civilization with the wherewithal for space travel should put all its eggs in one planetary basket. But somehow, the idea of a terraformed Mars seems unlikely. When you look at schedule for terraforming – first beefing up the atmosphere, then the introduction of hardy

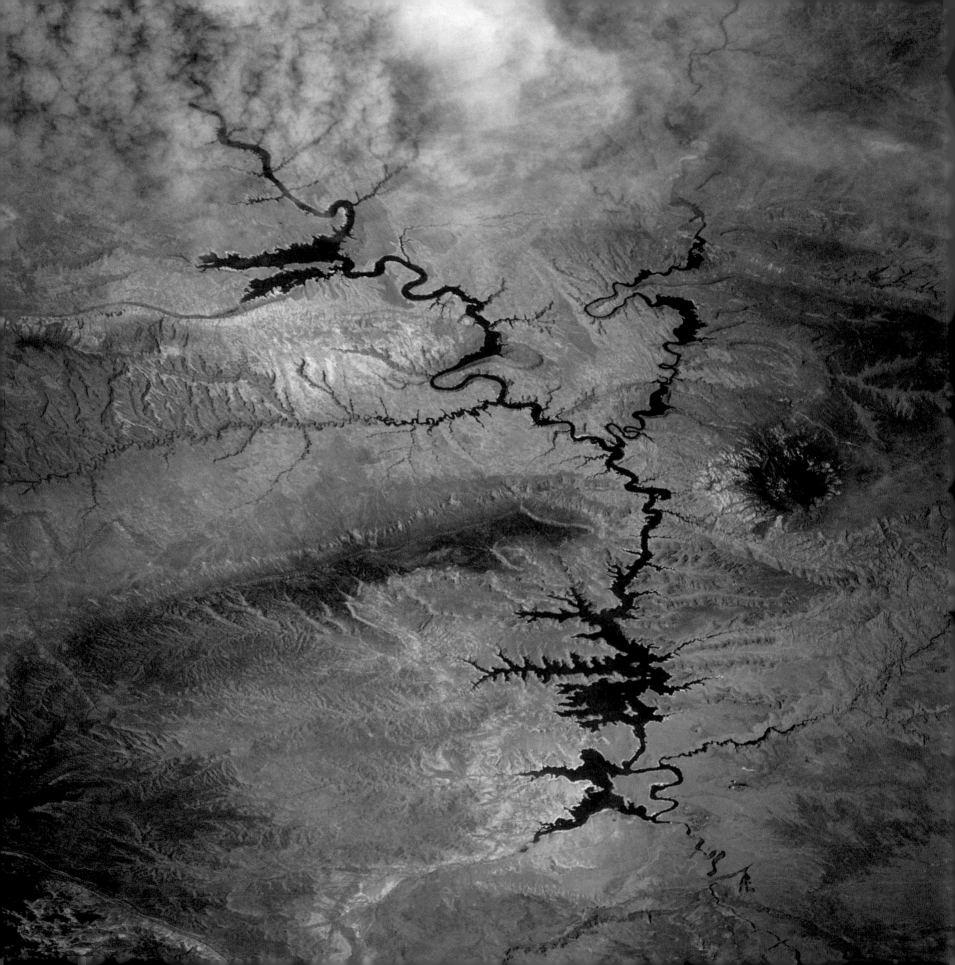

microbes, then larger organisms, maybe some lichens, proper plants, and then, finally, animals – what you are looking at is almost a bizarre recapitulation of the 'Fall of Mars', from Lowell to the Mariners, in reverse. It seems almost as if the terraformers are saying, 'What a pity Lowell was wrong, and the canals were an illusion – let's make them real!' By terraforming Mars we would be drawing the canals back onto its surface. It can be no coincidence that the idea of the New Frontier is very much an American one. Mars, after all, looks much like Arizona, and there are an extraordinary number of men and women researching the Red Planet who live and work in that state. In reality, it is almost impossible to contemplate the sequence of events necessary to terraform Mars happening on any feasible timescale. It would cost trillions, quadrillions, and take centuries, millennia even. Humans have managed to gain only the most tentative foothold on the continent of Antarctica, a place where you can breathe the air and have supplies brought in by steamer. Antarctica is a much easier place to live than Mars, yet there has been no effort to colonize it. Then there is the matter of morality. Mars may not have any intelligent beings, but it may have microbes, and colonization by terrestrial life-forms may spell the end for them. If life *is* detected on Mars anytime soon, then the environmentalists will probably argue that the planet should be kept in quarantine, forever off limits. Conversely, if another decade of robotic exploration fails to find anything alive, then the same people will probably argue that Mars should be left pristine. Either way, there will be strong (and not entirely irrational) arguments for being careful about establishing a human presence on the planet.

It is hard to imagine a future for humanity in which we do not eventually explore and even colonize other planets and moons in the solar system. Barring the collapse of our technological civilization, this may be inevitable. Humans have always tended to go forth, exploring and pushing back the boundaries of what is possible. Civilizations that were introverted, that saw no need to look over the horizon, have become moribund. In the Middle Ages the great Chinese Admiral Zheng He undertook a series of great voyages in a fleet of titanic junks. Some claim he even reached the Americas. If China had maintained its interest in seafaring, who knows what kind of world we would be living in today? In the end, an emperor declared such expeditions folly, and it was left to the Europeans to conquer the world. If Mars is to be conquered, it will be in the same tradition as the discovery and conquest of the New World, a conquest in the European tradition. It could be argued that there is a strong imperative for us to go to Mars as soon as we can. We (almost) have the technology, and it is not safe to assume that this will always be the case. Even if our planet is not hit by an asteroid, our civilization could collapse for any one of a hundred reasons, including the sort of environmental catastrophes being forewarned by the Green movement. After all, on a couple of occasions during the Cold War the world came quite close to a conflict which could have ended our modern world in a week. We may be looking at a very narrow window of opportunity – wait another generation or two, and we may be too busy surviving to worry about space exploration.

But people exploring and even living on Mars is not the same as the creation of a new Earth, a Terranova. Such an exercise will be so fantastically expensive, so fantastically time consuming, that its advocacy seems to ignore one of the realities of human nature – our inability and unwillingness to contribute to projects that will come to fruition only long after we are dead. Perhaps the last people to take such a long view were the rulers of ancient Egypt, whose colossal architecture was as grandiose and as out of scale with the lives of the common man as a project to terraform Mars would be today. Humans will almost certainly visit Mars, even live there. But it would take the will of the pharaohs to remodel it into our new home.

(*Opposite*) Just add water. This series of meandering channels could be on Mars – save for the blue stuff that is filling them up. This is not one of Kees Veenenbos's renderings, but a view of the meandering Colorado river as it winds its way across the arid terrain of southwest Utah. The picture is a reminder that from space Mars and Earth often look surprisingly similar, yet on the ground the two planets have been evolving in entirely different ways for billions of years.

Chapter 14
First Steps

'Our economy can certainly afford an effort of this magnitude'

Neil Armstrong (March 2004)

Since December 1972, humankind has been stuck in low-Earth orbit, and space exploration has been left to the robots. Although many assume that the Shuttle and Soyuz, the only existing spacecraft capable of carrying humans into orbit, are exploration vehicles, they are not. Neither can get beyond around 600 km (375 miles) from the Earth's surface – the distance, roughly, from Washington, DC to Boston, Massachusetts. In 1969 NASA had a vehicle – the Saturn V – which was capable of throwing nearly 120 tonnes into orbit. Today's biggest launch vehicles struggle to get 20 tonnes off the ground. Thirty years ago we seemed to be on the brink of developing a new propulsion system for deep-space missions – the nuclear thruster. Humankind would no longer be relying on a technology that was in essence more than a thousand years old, and would complete their conquest of the solar system in fast, sleek, atomic-powered, lightweight vehicles that could get to Mars, say, in a matter of weeks rather than months. It came to nothing.

Since the Nixon administration took that fateful decision back in 1969 to opt for the Shuttle/Space Station over any plans to develop either further lunar exploration or to send people to Mars, there have been numerous initiatives, plans, and schemes to get a human out of our planet's gravity well. The most ambitious, far-reaching, and plausible was the Space Exploration Initiative of 1989. That came to nothing, and those wanting with a passion to go to Mars had to wait. The 1990s were a frustrating time for space enthusiasts who wanted to see crewed flights to the planets. America had never been richer, nor more at peace, and yet the will just did not seem to be there. NASA was having difficulty raising money for unmanned missions, let alone crewed voyages, and its budget was dwindling in real terms year after year. The Reference Mission of the mid-1990s – a design concept for a Mars expedition incorporating elements of Mars Direct – was at least a working plan for how to go forward, but with no cash on the table it was just a paper dream.

The Russians did not give up on space exploration, but their Mars 96 probe was a terrible loss, and the glory days of the late 1970s and 1980s, when the Soviets became the first and, to date, only space power to manage to get something down in one piece on the infernal surface of Venus, were long behind it. By the start of the twenty-first century, a man on Mars never looked further away.

NEW HOPE: MARS EXPLORATION IN THE NEXT TWENTY YEARS

But during 2003 strange rumors started to surface. There was a report in the Los Angeles *Times* claiming, erroneously, that NASA's new administrator, Sean O'Keefe, had been authorized to develop a nuclear-powered rocket to take humans to the Red Planet. It turned out that Project Prometheus was rather less ambitious than that, involving (to begin with) only unmanned exploration, but Prometheus indicated that the nuclear option is again being taken seriously at NASA, and there seems to be a growing realization that it will be the power of the atom, rather than burning kerosene or hydrogen, that will get humans to the planets. In the shorter term, the Jupiter Icy Moons Orbiter (JIMO) is set to be the most impressive unmanned mission to date. The plan is for a 50 m (160 ft) spacecraft (half the size of the ISS) to fly to Jupiter and use its nuclear thermal propulsion

I BELIEVE IT IS HIGH TIME WE ACTED ON EVIDENCE OF MARTIAN W.M.D.

MISSION TO MOON, MARS

EXHIBIT A

EXHIBIT B

When President Bush made his announcement on 14 January 2004 that NASA was heading in a new direction – out of low Earth orbit, back to the Moon, and on towards Mars – his plans were treated with a degree of skepticism.

unit to make a grand tour of the moons Europa, Callisto, and Ganymede. The nuclear reactor on board will give JIMO unprecedented levels of power, not only for propulsion but also for its scientific instruments. Although a launch date for JIMO has been slipping in recent years, NASA hopes to name a prime contractor for the mission by 2005.

After the flurry of interest generated by Prometheus, there were further reports of a grand new initiative. In the run-up to the centenary of the Wright Brothers' first flight, which was to be marked by a large gathering at Kitty Hawk, reports again began to circulate that there was about to be a major announcement on manned spaceflight. Maybe President Bush was going to use the occasion to make a triumphant announcement – humans would return to the Moon in short order, and thence to Mars. It didn't happen. The celebration was a washout, the planned flight by a Wright *Flyer* replica crashed ignominiously, and Bush said nothing of interest.

But the rumors continued apace. On the night of 8 January 2004, a White House press officer briefed journalists aboard Air

Force One that a presidential announcement would be made the following week. The stories that appeared the next day were dramatic: the president would proclaim a return to the Moon by 2010, possibly even earlier. There would be a firm commitment to travel to Mars, by the end of the 2010s, the early 2020s at the latest. The Shuttle would be scrapped and a new, crewed spacecraft would be developed, a son-of-Apollo. There would be lots of new money for NASA – pundits spoke of at least a trillion dollars.

In retrospect, it was clear that this leak was a kite-flying exercise. There was a certain amount of skepticism about the Bush plans. As the rumors were building, Anne Applebaum, a columnist on the *Washington Post*, was scathing about any plans to revive manned space exploration. 'Mars, as a certain pop star once put it, isn't the kind of place where you'd want to raise your kids,' she wrote, 'nor is it the kind of place anybody is ever going to visit, as some of the NASA scientists know perfectly well.' Her argument, and many scientists agreed with her, was that manned exploration of Mars – or anywhere else for that matter – will always be an expensive, dangerous waste of time. Those with a vested interest in robotic space programs went on the airwaves to denounce any plans for reviving the old 1960s spirit at NASA.

Then came Bush's actual speech, delivered at NASA HQ in Washington, on the afternoon of 14 January. He outlined a near-future for NASA that was considerably less dramatic than the pundits had been predicting the previous week, but nevertheless represented a sea change for an agency that many say has simply lost its way. There would be no return to the Moon by 2010; this would come by 2013 at the earliest, probably more likely 2015. Nothing would really begin to happen until at least 2009. He stressed that the Moon would be the primary goal, with Mars to follow on. This was no 'Kennedy moment'. When JFK made his speech back in 1961, he was pledging to live with the consequences of an expensive new program right to the end of his presidency. He would receive the plaudits for Apollo's success, or he would carry the can for its failure. Bush was merely plotting a course for his successors – successors who could choose, as he well knew, to throw the whole thing out.

There was no specific commitment to Mars when it came to a date, nor was there an immediate commitment to increase NASA's budget, save for a one-off seedcorn fund of an extra billion dollars, to be spread out over five years, and an order to expropriate tens of billions of dollars from existing NASA budgets for the new manned spaceflight initiative. Bush confirmed that the Shuttle would be returned to flight, but scrapped by 2010, by which time construction of the ISS should be complete. America's commitment to the International Space Station would continue but, after the ISS was 'finished' towards the end of this decade, would be refocused away from microgravity science and towards preparing for a trip to Mars:

Today I announce a new plan to explore space and extend a human presence across our solar system. We will begin the effort quickly, using existing programs and personnel. We'll make steady progress – one mission, one voyage, one landing at a time ... [we will] develop and test a new spacecraft, the Crew Exploration Vehicle, by 2008, and conduct the first manned mission no later than 2014 ... [we will] return to the Moon by 2020, as the launching point for missions beyond ... With the experience and knowledge gained on the Moon, we will then be ready to take the next steps of space exploration: human missions to Mars and to worlds beyond.

Probably the most significant thing about Bush's speech was the commitment to develop a new manned space vehicle, the CEV – the first since Apollo in the 1960s. The commitment to the Moon seems firm, Mars far less so. Most importantly perhaps, there is nothing in the speech that could be used to bind Bush's successors to his long-term plan. Critics pointed out almost immediately that up to then, Bush Jr had never shown any interest in space. For a start, despite being governor of Texas for fourteen years, he had never once visited the Johnson Space Center, the capital of America's manned spaceflight program. For the time being, no one seems to agree on what Bush's vision will look like in the metal, and everyone is mindful of the hubris of the document that resulted from the grand vision of the president's father, the 90-Day Study, which maintained:

Historians will note that the journey to expand human presence into the solar system began in earnest on July 20, 1989 ... when President George Bush announced his proposal for a long-range continuing commitment to a bold program of human exploration of the solar system.

PROJECT CONSTELLATION

John Connolly is a Johnson Space Center engineer who has taken part in just about every design study of manned trips to Mars prepared by NASA in the last twenty-five years. He has also been involved in most of the JPL robotic explorations of Mars over the past three decades. Connolly is a supporter of a manned Mars mission. He has gone on the record as stating that the cost of such an expedition would be of the order of $30–40 billion, about half the cost, in real terms, of the Apollo missions in the 1960s.

Connolly says that Bush's announcement certainly cannot be written off as mere election propaganda, as many critics have said. Instead, it is not so much a call to arms, like Kennedy's speech in 1961 (which had a definite date for a Moon landing), but a call for a change of philosophy for a space agency which has, in many respects, lost its way:

The Bush speech is the most definitive direction NASA has received since Kennedy's moon speech. It sets Mars as a further-term goal, but also recognizes that we'll need to get our exploring legs back by tackling the Moon again first. Many of us who have been around the human exploration study world for the past dozen years have vivid memories of the 1989 SEI and its failings. So we will apply these lessons learned – in hardware design, mission design, policy, etc. where needed.

Some have criticized the emphasis on returning to the Moon. First, we have been to the Moon, and spending a great deal of time and money going there again seems a waste. Secondly, and more importantly, if Mars is the real, albeit long-term goal, then using the Moon as a 'stepping stone' makes no sense. One solid reason to go to the Moon might be to make fuel for the Mars ship; it seems bizarre, but it could actually be cheaper to make fuel on the Moon (if sufficient raw materials can be found there) and dispatch it back to Earth orbit than to lift it off the Earth's surface.

And what of the new crewed spacecraft mentioned by Bush – Crew Exploration Vehicle? Is this supposed to be a ground-to-orbit replacement for the Shuttle, or a son-of-Apollo, a crewed module for deep-space exploration? And surely building a Mars ship is going to need a rocket that can lift into space far more

The $600 million Mars Reconnaissance Orbiter, which is due to arrive at Mars in early 2006.

than the current 20-tonne limit of the Arianes, Protons, and Shuttles? Connolly again:

> You will eventually need launchers with greater than 20-tonne capability to build exploration missions for Mars. If all you have is 20 tonnes, everything you build will look like ISS. But, we do have a larger launcher in use today. The STS [Space Transport System – the Shuttle plus its boosters] launches a 240,000 lb [110-tonne] orbiter, which could be replaced by cargo.

In other words, NASA could use the Shuttle launch system – its main engines, external tank, and solid-fuel launch boosters – as a basis for a new heavy lifter, either crewed or uncrewed. This concept has been floating around for some time, and is known as 'Shuttle C'. As for the CEV, Connolly says:

> What this will eventually be is still under debate. I see it as an Apollo capsule-type function. Mission duration is related to the volume you need to keep people happy and productive. A weeks-long lunar mission is considerably different than a years-long Mars mission, so I don't see us using the same vehicle for both. Could a derivative of the lunar vehicle fly to Mars? Maybe. Could you add extra volume? Maybe ... the Russians do this with their Soyuz vehicle – it has a small 'habitation' volume that is separate from the more critical crew compartment.

By March 2004, some flesh had been put on the bones of the CEV concept. NASA's fiscal year 2005 budget laid out the development plans for the CEV, and stated that the first unmanned demonstration flights would take place in 2008. Before the Bush speech, the Agency was facing a spending freeze that would have locked the annual budget, at least to 2009, at a level of some $15 billion annually. With its new mandate, NASA funding will grow steadily from around $16.2 billion in 2005 to $18 billion in 2009.

The overall strategy is to return to the Moon in one go and make sure that the overall project cannot be threatened by a single policy change. Project Constellation, as the initiative has been christened, will use small incremental steps called 'space policy building blocks' in which each blocks is a discrete mission or set of missions in its own right. The first of the blocks, called

Lunar Testbeds and Missions, will include a new series of robotic probes to the Moon, along with an accelerated program of unmanned Mars exploration – which may include a definite sample-and-return mission for 2011. NASA already has announced a new series of reconnaissance satellites to be launched into lunar orbit. Scheduled for 2008, these probes will comprehensively map the entire lunar surface, so that charts can be produced for the new generation of explorers. There will then be a robotic lunar lander in 2009, followed by more landings (robotic) and the establishment of a communications network.

The second building block is to be called Mars Research, Testbeds and Missions. Before humans can land, the surface of Mars must be mapped, surveyed, and analyzed, and all possible dangers accounted for. So far, four future missions have been fully funded and will definitely fly. Missions have been planned to take advantage of each 26-monthly launch window – the time at which a spacecraft can get to Mars in the minimum time and using the least amount of fuel.

The next JPL mission will be the $600 million Mars Reconnaissance Orbiter (MRO), scheduled for launch in August 2005. This will be a truly extraordinary machine, equipped with a camera that will be able to image objects as small as a coffee table, and fitted with a spectrometer that will be able to see from orbit the sort of individual sulfate deposits that Opportunity has seen on the ground. The telescopic camera, the High Resolution Imaging Science Experiment (HiRISE), will be the most powerful ever sent into the orbit of another planet. The 2-tonne (with fuel) MRO will also scan underground layers for water and ice, identify small patches of surface minerals to determine their composition and origins, track changes in atmospheric water and dust, and check global weather every day.

Mars Reconnaissance Orbiter (MRO) is very much a continuation of the follow-the-water strategy. With its penetrating radar, it will check whether the frozen water that Odyssey detected in the top meter or two of soil extends any deeper. According to MRO's project scientist Richard Zurek, we need far more detailed maps of Mars than we have now if we are to select the best landing sites. 'A key lesson from the MER missions is that it is still difficult to link what we see from orbit with what we see on the ground', he says. 'This pushes us to try for greater spatial resolution.' The MRO design should also overcome one of the great bugbears of Mars exploration to date – the communication problem. The spacecraft will communicate with Earth using the widest dish and most powerful transmitter ever used at Mars. According to James Graf, the MRO project manager, it will be like 'upgrading from a dial-up modem for your computer to a high-speed DSL connection'.

Two years after the launch of MRO comes Phoenix, the next lander scheduled for Mars. Phoenix is essentially a revived but improved version of the old Mars Surveyor Program 2001 'lander' that was cancelled in early 2000 after the failure of Mars Polar Lander in December 1999. As JPL's website more poetically puts it, 'like the Phoenix bird of ancient mythology, the Phoenix mission is reborn out of fire.' Phoenix will carry improved instruments, including a gas analysis package with a highly sensitive mass spectrometer capable of detecting minute quantities of organic materials – a far more sensitive instrument than that carried aboard the Vikings. It will also be equipped with a 'wet chemistry' lab, three microscopes, a large arm capable of digging a meter underground, and a Canadian meteorological package. It is due to land on the ice-rich northern plains between 65° and 75° latitude in May 2008 and is designed to operate for 90–100 days. Its primary mission is to 'search for evidence of a habitable zone and assess the biologic potential of the ice/soil boundary'.

Next off the launchpad will be the 2009 Mars Science Laboratory (MSL). This, in the words of on-line magazine marsnews.com, 'is an all-terrain, all-purpose machine, akin to an extraterrestrial Sport Utility Vehicle'. MSL, which is supposed to cost less than $850 million (launch not included), will be the biggest, most powerful, and most advanced lander in space exploration history. The final design has yet, at the time of writing, to be decided, but one leading candidate introduces a completely novel and elegant way of getting down onto the surface of Mars – the Skycrane. After a normal descent slowed by parachute, MSL is brought to the surface tethered under a large gantry arrangement which is equipped with powerful retro-rockets to slow its descent to a standstill 5 m (16 ft) or so above the surface. The whole structure hovers in this position for as long as 5 seconds (using artificial-intelligence software to adjust its position to account for winds – no real-time control will be possible from Earth, because of the time delay), during which time the rover slips down the tether to the surface of Mars. Then, its work done, the Skycrane flies off and crash-lands out of harm's way, several hundred meters away from the arrival area. The advantages are clear – no bouncing, no rolling

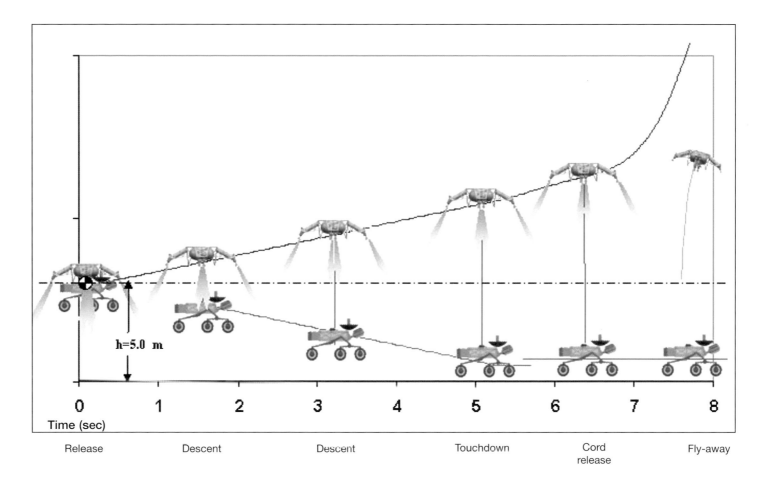

h=5.0 m

| 0 | 1 | 2 | 3 | 4 | 5 | 6 | 7 | 8 |

Time (sec)

Release · Descent · Descent · Touchdown · Cord release · Fly-away

The Skycrane is the most innovative – and complex – landing system yet designed for a planetary lander. If approved, the Mars Science Laboratory landing system will consist of a rocket-powered gantry that will gently lower the rover to the surface, avoiding the randomness of an airbag landing.

down crater walls or getting wedged under rocks, no periods of radio silence. Best of all, the final touchdown speed is just 1 m (3 ft) per second – no 20g jolt to confuse software or damage hardware.

The current rover–lander design is five times larger than the MERs. In its size and complexity MSL more resembles an enlarged and updated mobile Viking lander, weighing almost a tonne, with ten times the science payload of the latest landers. One key difference would be a return to nuclear power – MSL will be the first nuclear-powered spacecraft to land on Mars for thirty-four years, and it is understood that both Boeing and Lockheed Martin are currently working on competing designs for the nuclear batteries that will power MSL for more than 600 days on the Martian surface.

By 2009, images and movie footage from MSL will be released in real time over the Internet, and a hopefully successful mission will entertain millions. The landing site has not been decided upon, and will depend on data returned by the current landers, Phoenix, and the Mars Reconnaissance Orbiter. There is talk of sending MSL to one of the more 'exciting' places on Mars, maybe even into the Valles Marineris, which has so far been strictly off-limits because of the ruggedness of its terrain. Meridiani is another possibility, as it is turning out to be rather more inter-esting than anyone dared hope. MSL will take advantage of the Telecommunications Orbiter, also due for launch in 2009, which will provide a high-bandwidth relay station for communication with Earth. Over the coming six years, Mars will be slowly wired up to the terrestrial communications system.

THE NEXT DECADE: 2010-2020

What happens after MSL depends a lot on what is discovered in the next six years. Nothing is set in stone, nothing has been funded for sure, but according to Jim Garvin one of several possible 'pathways' will be chosen for Mars exploration depending on the results of current and near-future discoveries. Should evidence of extant liquid surface water be found, then that will obviously determine the direction of the whole program, as would the discovery of any 'live' geothermal activity. What is certain, Garvin says, is that NASA will 'search aggressively for past life'.

The key mission of the second decade will be a $3 billion sample-and-return mission, planned provisionally for 2013. Getting a piece of Mars back to Earth has been something of a holy grail for planetary scientists for decades. Meteorites have been found that are known to have come from Mars, but they have been exposed to the terrestrial environment for perhaps centuries and may also be highly atypical, because of they way they were ejected from the Martian surface in the first place. What scientists would really like to get their hands on is a fresh, unaltered chunk of Mars rock or soil which they can study under sterile conditions. Also, a sample-and-return mission would very much be seen as a precursor mission to a human landing, as it would need to incorporate a vehicle to lift off from the surface of Mars and return to Earth. There is a plan to send, by 2016, a dedicated life-search mission – the Astrobiology Field Laboratory (AFL), which will have at its disposal incredibly sensitive life-detection experiments. Peering further into the fog of the crystal ball, there are also plans on the drawing board for a deep-drilling mission to search for biological material at a depth of perhaps 30 m (100 ft) or more. This may be an and/or mission with respect to the Field Laboratory, depending on what previous missions establish and on the funding available towards the end of the 2010s. By the end of the next decade, all being well, we should know enough about Mars to start making realistic plans for a crewed mission.

The first crewed lunar landings of the new millennium are due to take place around 2015, about the same time as the first robots designed to 'mine' the Martian atmosphere for rocket fuel and dig for water are launched. At this point, Project Constellation will merge with Project Prometheus. All this will pave the way for a triumphant Mars landing sometime in the 2020s.

This is what is known about the 'new direction' at the time of writing. But Connolly and others at NASA say that it is far too early to make any predictions about what form the end result – a crewed Mars mission – could take. Maybe they would take on board some of the concepts developed by Bob Zubrin and his Mars Direct team. 'We give Bob credit for positively influencing us,' Connolly says, citing certain elements of a 1993 design reference mission that utilized many aspects of Mars Direct. Connolly concludes that:

> Returning to the Moon between 2015 and 2020 is certainly do-able if the resources and support are in place. NASA and others have done studies in the past that guesstimate the period of time required to learn lessons from lunar missions, and then apply them to Mars.

THE INTERNATIONAL DIMENSION

An analysis that appeared in the Russian journal *Novosti* in January 2004 gave some insight into Russian thinking on all this:

> In his January 14 speech, President George Bush unveiled a truly fantastic national space program which, to all appearances, would seem to involve the United States and the entire international community in further space exploration. Analysts openly note that this program can only succeed if the White House opts for wide-ranging cooperation with Russia. Moscow agrees with this approach completely.

Interestingly, when the Bush plans were first leaked, there was much talk of international cooperation. Sources inside the Russian Space Agency point out that NASA has approached Moscow on more than one occasion to talk about the Moon and Mars. Russia has unique experience in long-duration spaceflight, and the data they have acquired on Mir, Salyut, and so on will be invaluable in planning any Mars mission. If men and women are going to spend months on end cooped up in a tin can on a voyage incomparably more hazardous (if incomparably more comfortable) than those undertaken by the mariners who mapped the world in the sixteenth and seventeenth centuries, then the mission planners must get the psychology right. And there is good evidence that, when it comes to understanding how a crew works as a group of people rather than a group of pilots, the Russians are rather more knowledgeable than their American counterparts. Mike Foale, the British-born astronaut who is now NASA's leading astronaut, visited Mir in 1997. He says how important it was for him to fit in with the Russian space ethos, which is as much about sociability and working as a team as it is about technical expertise. On the voyage to Mars, cooking expertise, playing a guitar, or simply knowing when to shut up will be just as important as the abilities to pilot a lander, repair an air-conditioning unit, or identify an olivine basalt when you see one. The Russians have long experience with crews of varying makeup, and indeed have sent more than one husband-and-wife couple into orbit together.

In 2001, the European Space Agency member states also agreed on a program that would 'maintain an interest in human spaceflight after the completion of the International Space Station', with the focus and eventual goal being a crewed landing on Mars. Like the current NASA plan, Aurora, as the program is known, will feature a series of robotic missions to Mars – and other bodies, such as the Moon and asteroids – followed by crewed exploration. Aurora has already drawn a highly enthusiastic response from the scientific community, with dozens of proposals being put forward for missions and specific science packages. ESA plans a sample-and-return mission between 2011 and 2014, bringing back about 0.5 kg (1 lb) of material. The stated goal is a crewed landing sometime around 2033.

Aurora is an ambitious 'roadmap' but at present it remains little more than that, as at the time of writing there was no definite funding in place for any missions beyond 2005. The first new mission, ExoMars (a sophisticated lander/rover) was due to get a funding decision in late 2004. The most ambitious element of Aurora, in the short-to-medium term, is a sample-and-return mission planned for launch in 2011, with a return of Martian material in 2013. If ESA can get this off the ground, they could beat their rivals in Pasadena by several years.

However, many of the Aurora objectives seem to duplicate those of the NASA Mars program. Would it not be sensible for

Project Aurora is Europe's vision for Mars exploration. It includes rovers, sample-and-retrieve missions, and even a crewed landing in the 2030s. Whether the reality of Aurora will match up to the vision is a matter for the bean-counters.

the Europeans and the Americans to pool resources? Jim Garvin says that there are funding issues to resolve:

We all agree [that we should cooperate] and many ESA Mars scientists want us to work together. I think the brilliant success of Mars Express and the MERs offers a good example of how we can work together well. But at issue are the funding realities involved. We are spending up to $600 million a year on Mars and are already spending many tens of millions on Mars Science Laboratory and Phoenix. Whenever I talk with the leaders of Aurora about budgets ... we receive limited information, and the suggestion is that there are very few euros to start building the system and funding the instruments . . .

Mars is out there awaiting us ... but as we have seen, it is also very unforgiving. When it comes to humans, we have to learn how to do this safely. [If] Columbus . . . hadn't come back, would we have remembered him?

Whatever the uncertainties over design and funding, there is a real sense that Mars exploration has entered a new era and is gathering pace. There will undoubtedly be more failures and heartbreaks at mission control. We probably couldn't go to Mars right now if we tried, but it is likely that in thirty years time the situation will have changed. With new resources and new technologies, such as nuclear propulsion, the Red Planet may come within our reach.

PRESSING THE CASE FOR HUMAN EXPLORATION

It is important to remember the current Bush proposal is a plan for the future of NASA, not a plan for an expedition to Mars *per se*. NASA and the other space agencies still need to work out how they are going to sell the idea of human Mars exploration. Many scientists are resolutely opposed to the idea, claiming that robots can do the job just as well, if not better, for a fraction of the cost. However, it is hard to argue this case very convincingly. Robots are cheaper than crewed spacecraft, but they have severe limitations. If a robot lander's batteries fail to recharge, it is dead. If an astronaut gets hungry, he eats a sandwich. Humans are superb, adaptable, all-purpose decision-making machines. Mars is always several light-minutes away, and running a mission on its surface in true real time is therefore impossible. Despite the undoubted advances in computing technology, real artificial intelligence is till an oxymoron. Sojourner never strayed more than a few meters from its base station; Spirit and Opportunity are far more independent and may venture a few kilometers; but Opportunity, for example, spent the first month of its mission in the crater in which it had landed. Wonderful though the work it did there was, there is little doubt that the same tasks could have been performed by a human in an hour. That is why human explorers will be needed if we are to ever unlock the secrets of the real Mars.

The case for Mars must be made partly in the spirit of exploration, and mostly in the spirit of science. Sending people to Mars is dangerous and hugely expensive, but it could also be extremely profitable. A common refrain from those wanting to send people to Mars is that while it may cost a hundred times as much to send a human to Mars than to send a robot, he or she can accomplish a thousand times as much science. The opponents of crewed missions tend to massively underestimate the usefulness of humans. Jim Garvin, head of Mars science programs at NASA, is enthusiastic about sending people to Mars – so much so that in 1976 he volunteered only half-jokingly for a one-way ticket to Mars if that was the only way of getting a scientist onto its surface. 'I wouldn't do that now – I have children and a dog to worry about', he says. Garvin believes that sending people is an incredibly cost-effective way of exploring planets:

> None of us doubt the incredible value of having humans there. One of the holy grails of the Mars program is to return pieces of Martian material. The best we can imagine using robotic probes over the next twenty years is the return of a little less than one kilogram of the stuff – maybe only half a kilo – at a cost maybe of $2–3 billion. One human mission for thirty days could return a hundred times the mass of that sample. Look at the hundreds of pounds of Apollo samples that we are still studying.

But, of course, the science will not be enough for the public. They will want drama and glamour – amazingly, in Apollo, NASA managed to make exploring another world seem rather *boring*. If NASA does indeed send a rocket much farther than a crewed vessel has ever traveled before, sometime in the late 2020s or early 2030s, the people on board will instantly become global megastars. Their lives, loves, and habits will be on primetime TV (or whatever technology has superseded it), twenty-four hours a day, seven days a week. The Apollo missions each lasted a week or so. The Mars

(*Opposite*) Both ESA and NASA have plans for a sample-and-return mission from Mars, sometime in the 2010s. Returning a sample of Martian soil or rock would be a boon to geologists, but it would be expensive – around $3 billion for a kilogram of material. This has led some scientists to question whether it may, in the long term, be more cost-effective to send humans. Much of the value of the Apollo program came from the several hundred kilograms of lunar rocks returned by the astronauts.

If the light is bright enough, the so called 'range of entrainment' by the clock can be expanded and would probably accommodate a slightly shorter or longer period than twenty-four hours. But you can't push it too far. I doubt that we could manage much beyond twenty-five hours or less than twenty-three. My feeling is that on Mars using natural light they may be able to lock on. So windows would be a good thing, as would spending time outside. The trouble is, if you want to use Martian light there are problems with ultraviolet radiation and so on.

We already know what living on Mars time does to people, and not just from the experiments performed by the NSBRI. At time of writing, the teams at JPL running the Mars Exploration Rovers had been on Mars time for weeks and were absolutely exhausted. If the astronauts do adapt to Mars time, it would probably be a good idea if mission control did the same. 'The problem is, if something goes wrong on Mars then you want the people on Earth to be absolutely fine-tuned', says Foster.

He may be overstating the problems of 'Mars time': many humans can – and do – live quite happily by a whole variety of strange clocks. Not everyone at JPL who has been living on Mars time has had problems – these have arisen only when they are forced to interact with the real world. No one knows for sure, but the suspicion must be that an ape which evolved in tropical Africa and then went on to colonize the world will be able to cope with Mars, however alien it at first appears.

We have been talking about going to Mars for nearly a century now. Are we any closer than we were, say, forty years ago? Probably. With hindsight, it is clear that the technological challenges and human dangers of going to Mars were hugely underestimated in the heady days of Apollo. Technology has progressed tremendously, yet NASA's planners and engineers seem to have even less of an idea than did Wernher von Braun of how to get a group of people to Mars and back again safely. We are more realistic now, though we have lost the spirit that was generated by the relentless optimism of the 1960s, and by the Cold War patriotism that liberated the cash. Maybe Apollo was an anomaly, a program that came forty years too early.

The history of going to Mars (or rather, the history of not going to Mars) has parallels with our changing perception of the planet itself. A hundred years ago we knew that Mars had a thick atmosphere, was warm and wet like the Earth, and was probably peopled by canal-building aliens. Now we just know that we know very little. Fifty years ago, getting to Mars was easy; now we know that it is very, very hard. Thanks to the discoveries at Mars, there finally seems to be a momentum building up, a feeling that serious exploration is under way, and that soon we will need to go and see these channels and layered rocks in person. Only time will tell if this is a realistic dream or whether it evaporates into what-ifs and science fiction, like all other attempts to go to Mars.

(*Opposite*) If and when human explorers eventually do locate the Viking landers, the layer of dust that will have built up on them may be considerably thicker than depicted here.

Conclusion
The Real Mars

'We are all the consciousness that Mars has ever had'

Kim Stanley Robinson (*Red Mars*, 1993)

What is the real Mars? We know this planet from our telescopes and our space probes, from the musings of writers and astronomers, and from bits of rock that have been blasted across the void to end up on some Antarctic snowfield or on an Egyptian dune. Our Mars is a planet that is constantly changing, a place where life appears and disappears in a lifetime, shaped not only by science but by our imaginations. Our Mars is an incomplete Mars, is a place of uncertainty. Are there volcanic rumblings still under the Tharsis bulge? Was Mars ever once warm and wet, a blue planet with an immense ocean covering the north, and gushing rivers longer than the Nile feeding in from the southern highlands? Does water, to this day, flow down crater walls, trickles of frigid brine glinting in the afternoon sunlight? Might microbes still lurk around geothermal springs? Could pores in rocks deep underground be home to tiny beasts? Will Mars ever be our second home?

The old Mars was a vague sort of place, *terra incognita* until the great pseudo-discovery of the canals by Percival Lowell. It then became a world of desert kingdoms and strange creatures. But to some, Mars is as alive today as it has ever been. Where once there were warriors there are now ancient pyramid-builders, beings who send messages across space with their carved sphinxes, and who built giant cities among the jumbled rocks of Cydonia. To others, Mars is a world waiting to be conquered, a place where human history can be reborn, untainted by the mistakes of the past. The Mars colonists see a planet re-made in the image of liberal America, a new Wild West. It is no coincidence that the new Martian frontier so resembles the older American frontier, both in appearance and in spirit. To the terraformers and colonists-in-waiting, Mars,

like Arizona, is a red, harsh world that can nevertheless be tamed. The deserts will one day bloom, watered by rains falling from skies altered by the works of man. Mars, like the Wild West, is both unforgiving and welcoming.

Mars is constantly changing, partly because we are learning more and more about it, but also because what we want has changed. For most of the twentieth century we wanted to believe there was life on the fourth planet, perhaps even higher forms of life. It is surprising how readily the idea of Martians was accepted by writers and by scientists. As late as the 1960s, Carl Sagan was still talking about Martian 'polar bears'. When he first considered sending men to Mars, Wernher von Braun thought that the landing party should be armed, in case it ran into hostile fauna. More realistically, a few years later Patrick Moore could not rule out 'a few lichens'. No one wanted to think about Martian life being restricted to perhaps a few hardy colonies of bacteria. Mars fell to perhaps its lowest ebb after the Mariner 4 flyby showed craters, dust, and a near-vacuum of an atmosphere. In fact, what Mariner saw was neither wholly typical of the Martian surface, nor should it have been totally unexpected. A lack of craters would have been far more surprising, and estimates of the density of the Martian atmosphere had been falling for years.

Mars started to change again in the late 1990s. The exploration of the planet suffered many setbacks as the twentieth

(*Opposite*) The famous imagery of the Reull Vallis by Mars Express is striking, but the unnaturally vivid colors – especially the 'watery' bluish black of the valley floor – may simply reveal the strength of our desire to 'see' vast river basins and drainage channels.

century drew to a close, with a whole series of probes falling foul of the 'Mars Ghoul'. But the late 1990s were also a time when Mars really started to come back to life. Mars Global Surveyor has been an outstanding success. Its brilliant camera, operated by Mike Malin and his team, has repainted the surface of this world. We now have better maps of Mars – taking its surface as a whole – than we do of Earth, which of course is mostly covered by water. It was images of gullies from this camera in 2000 that suggested there may be life in the old planet yet. What were those gullies? Were they drainage channels carved by trickling streams of salty water, or perhaps they were made by ice?

Percival Lowell would have given all his limbs for the views we are getting of Mars today. Through his telescope the planet appeared no larger than a dime held at arm's length. What would he have made of the gullies, the channels, the layered deposits in the craters? How would Lowell have interpreted the Martian Grand Canyon, the strange pits in the southern polar cap? Where would Lowell have stood on the warm, wet Mars controversy? Would he have seen a face in Cydonia, or just a pile of rocks?

We seem to have come full circle. Back in the 1890s, Mars was a dynamic planet, a place of shifting vegetation, canals, polar seas, and strange lilac flashes in its atmosphere. For the next ninety years Mars slowly desiccated and died. Now it is dynamic again. Its volcanoes may be dormant or dead, its rivers dried up, and its Martians just a fantasy, but here and there are hints of movement – trickles of water, shifting sands. When the meteorite fossils crawled onto the front pages in 1996, here was a hint that Mars's past may be even more exciting than its present. Liquid water has a hard time on the Martian surface of today, but *something* must have carved those channels.

John Brandenburg and others imagine a planet not unlike a primordial Earth, with a dusting of green where now is only red. Perhaps there were Martians, real Martians – not men, but fish and slithering things – which lived more than 3.5 billion years ago. This is an extreme view, but what is

becoming generally accepted now is that Mars did once have a lot of water. There are those river channels, for a start. And the topography of Mars suggests, to some, that its surface was once covered with much larger bodies of water – lakes, seas, maybe even a great ocean, the Martian Oceanis Borealis, whose waves lapped for maybe a million centuries upon an ancient shoreline that can still be glimpsed today. The laser altimeter instrument aboard Mars Global Surveyor has produced stunning relief maps of the planet. The fact that the scientists have chosen the color blue to represent the lowest elevations has brought the ghost of that ancient, watery Mars swimming back into view, with great splurges of azure suggesting ancient seas. Most Mars researchers today tend to hold the opinion that Mars must have had warm and wet periods, even if these were only brief interludes in an otherwise frigid history. The findings from Opportunity, which spent the early part of 2004 trundling around the flatlands of Meridiani, support the idea of a Mars that was once damper and warmer than today.

New Mars, new adventures. It is at least possible that had those early Mariners taken photographs of Lowell's Mars, we would now be a species inhabiting two worlds. There would have been a manned Mars program by the 1970s or 1980s. But this was not to be. The Mars of the Mariners was too 'dull' to justify spending the billions that such a mission would cost. Now things may be turning. NASA has found in Mars the sort of jewel that it had lost with Apollo – a place of adventure and discovery that calls for bold, new missions. Maybe in the coming decades we shall see the revival of those dreams to send people to the Red Planet. The first crewed Mars mission will be nothing like the gargantuan exercise planned by Wernher von Braun and the other space pioneers. It will be small and lightweight, fast and international. It seems unlikely that there will be a human on Mars much before the middle of this century. It won't happen before 2025, probably 2030. But there are now grandiose plans to revive America's manned space program. In early 2004, President Bush announced a plan to return to the Moon, and thence to Mars. Yet such a mission is looking more likely than for decades.

The Real Mars remains a mystery. We haven't even scratched its surface. It will no doubt contain wonders which we can only imagine; it may be home to life, or it may be not. One day, probably, it will be visited by humans. But even if and

(*Opposite*) This rendered view of the Ophir and Candor Chasmae, part of the Valles Marineris canyon complex, shows Mars's most dramatic scenery as it appears today.

when humans visit the Red Planet, it will take us a long time to get to know this new world.

It is easy to regard Lowell today as a figure of fun – a man who peered through a telescope and saw what he wanted to see. Who is to say that, a century hence, our descendents will not be making the same judgments about us, with our gullies and microbes, ancient riverbeds and underground springs? After all, we, like Lowell, are only relying on the evidence of our own eyes. Where some see vast river basins and drainage channels, others see wind-scoured volcanic dykes or exotic features made by carbon dioxide. Just because we have a high-resolution photograph taken by an orbiting spacecraft doesn't end the debate any more than, in 1905, the first decent photographic images of Mars from the Lowell Observatory were put on display. When they saw the pictures, the scientists were divided; some saw canals, others did not. Lowell himself freely conceded that one needed to be well acquainted with telescopic photography to 'see' what he saw as clear as day.

Why were the early Mariner findings referred to as the 'Great Disappointment'? What is innately bad about a cold, dry, dead Mars? The Sahara Desert is drier and more lifeless than the Amazon jungle, yet neither place is judged to be 'better' than the other. Even though Lowell's canals were shown to be false, the problems, controversies, and intrigues of Mars have been discussed within a Lowellian framework and in Lowellian terms since his death. Lowell has become something of a figure of fun in scientific circles, but that is to do him a grave disservice. Lowell was one of the most talented observers and meticulous collectors of data of his generation. Although he was wrong about the canals, he was right about much else. He and his colleagues did much to advance the cause of astronomy in the early twentieth century, discovering, among other things, ways of measuring the atmospheric composition of the planets using spectroscopy, and of course it was at the Lowell Observatory that 'Planet X' – Pluto – was discovered in 1930, as a result of a search instigated by Lowell himself. Lowell's Mars is still the dominant Mars in the popular imagination of today, even if most people know that canals and Martians are a fantasy.

Lowell's Mars gave rise to Lowell's universe, a universe in which Earth is not alone, but probably populated by teeming hordes of aliens just waiting to be discovered.

William Hoyt concluded his book *Lowell and Mars* with this summary:

> If Lowell's canal-building Martians are today in total disrepute, the possibility of extraterrestrial life which was the fundamental premise of his controversial Mars theory has become not only widely accepted in science but has proved a rationale for a number of highly imaginative investigations which, if far more sophisticated, are hardly less 'fanciful' than the one that Lowell embarked on at the turn of the twentieth century.

Lowell may be credited, with H.G. Wells, with inventing the whole concept of the alien. Before the 1890s, Martians (and the inhabitants of other planets) were usually conceived of as being simply human, or perhaps variations on the human theme. After Lowell, the Martians became superhuman. He emphasized repeatedly the vast intelligence and the efficient oligarchic society without which it would have been impossible to construct the huge canal network. In dozens of movies and hundreds of books and short stories, the 'Martian' is invariably cleverer, better armed, and equipped with superior technology and/or sensory apparatus than his human counterparts. It is hard to think of a single example from science fiction where Martians are portrayed as dim-witted, simple-minded, or primitive compared with the inhabitants of this planet.

When the announcement of the Martian bugs in that meteorite was made, someone left a note and a glass of champagne outside Percival Lowell's mausoleum, just across the way from the Clark Telescope. 'Percy,' the unsigned note said, 'you were right all along.' Lowell's view of Mars survives to this day. The canals may have vanished into the dry sands of Mars as soon as we looked up close, but squint your eyes a bit, say the skeptics, look away from that eyepiece, and maybe they are still there.

Appendix
Mars Missions

PAST MISSIONS

Missions	Date	Country
Marsniks 1 and 2	1960	USSR
Sputnik 22	1962	USSR
Mars 1	1962	USSR
Sputnik 24	1963	USSR
Mariner 4	1964	USA
Mariner 6	Past1969	USA
Mariner 7	1969	USA
Mariner 9	1969	USA
Mars 2 and 3 landers	1971	USSR
Viking orbiters and landers	1976	USA
Phobos 1 and 2	1988	USSR
Mars Observer Mission	1993	USA
Mars Global Surveyor	1996	USA
Mars Pathfinder	1996	USA
Mars 96	1996	Russia
Nozomi probe	1999	Japan
Mars Climate orbiter	1999	USA
Mars Polar Lander	1999	USA
Deep Space 2	1999	USA
Mars Odyssey	2001	USA
Beagle 2	2003	Britain
Mars Express	2003	EU
Mars Exploration Rovers	2004	USA

FUTURE MISSIONS

Mission	Date	Country
Mars Reconnaisance Orbiter	2005	USA
Phoenix Lander	2007	USA
Mars Science Laboratory	2009	USA
Aurora	2011	EU
Astrobiology Field Laboratory	2016	USA

Bibliography and References

The following list includes those major sources that are quoted from, or otherwise used, in the text. Where quotes have not been attributed they resulted from conversations or correspondence with the author.

BOOKS

Austin, A.B., *The Sand Whales of Mars*. Black Rabbit Press, 2001.

Bradbury, Ray, *The Martian Chronicles*. Grafton, 1951.

Brandenburg, J.E. *et al.*, *Dead Mars Dying Earth*. Crossing Press, 2000.

Carr, Michael H., *Water on Mars*. Oxford University Press, 1996.

Davidson, K., *Carl Sagan. A Life*. John Wiley & Sons, 1999.

Davies, P., *The Fifth Miracle: The Search for the Origin and Meaning of Life*. Touchstone Books, 2000.

Dyson, G., *Project Orion: The Atomic Spaceship*. Allen Lane, 2002.

Flammarion, C., *La Planète Mars*. Gauthier-Villiers et Fils, 1892.

Grady, M., *Search for Life*. The Natural History Museum, London, 2001.

Harland, David M., *The Earth in Context*. Springer Praxis, 2001.

Hartmann, William K., *A Traveler's Guide to Mars*. Workman Publishing, 2003.

Hoagland, R., *Monuments of Mars: A City on the Edge of Forever*. North Atlantic Books, 1992.

Hoyt, W.G., *Lowell and Mars*. University of Arizona Press, 1976.

Jones, D., *Horror, A Thematic History in Fiction and Film*. Arnold, 2003.

Kant, I., *Universal Natural History and Theory of Heaven*. 1755.

Leonard, G.H., *Someone Else is on Our Moon*. W.H. Allen, 1977.

Lowell, P., *Mars*. Longmans, Green & Co., 1896.

Lowell, P., *Mars and Its Canals*. Macmillan Co., 1906.

Lowell, P., *Mars as the Abode of Life*. Macmillan Co., 1909.

Moore, Patrick, *Guide to Mars*. Frederick Muller, 1956.

Moore, Patrick, *Guide to Mars*. Frederick Muller, 1965.

Moore, Patrick, *Guide to Mars*. Lutterworth Press, 1977.

Moore, Patrick, *Patrick Moore on Mars*. Cassell, 1998.

Morton, Oliver, *Mapping Mars*. Fourth Estate, 2002.

Pickering, W., *Mars*. Gorham Press, 1921.

Pillinger, C.T. *et al.*, *The Guide to Beagle 2*. C.T. Pillinger/Faber & Faber, 2003.

Piszkiewicz, D., *Wernher von Braun: The Man Who Sold the Moon*. Praeger, 1998.

Pohl, Frederik, *Man Plus*. Victor Gollancz, 1976.

Robinson, K.S., *Red Mars; Green Mars; Blue Mars*. Voyager, 1992–1996.

Slipher, E.C., *Mars: The Photographic Story*. Northland Press, 1962.

Turner, M.J.L., *Expedition Mars*. Springer Praxis, 2004.

Vaucouleurs, Gerard de (transl. Patrick Moore)., *The Planet Mars (Le Probleme Martien)*. Faber & Faber, 1950.

von Braun, W., *The Mars Project*. University of Illinois Press, 1953.

Wallace, A.R., *Is Mars Habitable? A Critical Examination of Professor Percival Lowell's Book, Mars and its Canals, With an Alternative Explanation*. Macmillan, 1907.

Zubrin, R. and Wagner, R., *The Case for Mars: The Plan to Settle the Red Planet and Why We Must*. Free Press, 1996.

ARTICLES AND MONOGRAPHS

Historical observations and cultural history

Clerke, A., 'New views about Mars'. *Edinburgh Review*, vol. 184, 1896, pp. 368–84.

Crossley, R., 'Percival Lowell and the history of Mars'. *The Massachusetts Review*, vol. 41, no. 3, Spring 2000, pp. 297–318.

Lowell, P., 'The habitability of Mars'. *Nature*, vol. 77, 1908.

Solomon, S.C., 'An older face for Mars'. *Nature*, vol. 418, 2002, pp. 27–8.

Wells. H.G., 'Intelligence on Mars'. *The Saturday Review*, 4 April 1896.

Letters and documents from the Percival Lowell archive held at the Library of the Lowell Observatory, Flagstaff, Arizona.

Geology, water, ice and climate change on Mars

Ahronson, O. *et al.*, 'Slope streaks on Mars: correlations with surface properties and the potential role of water'. *Geophysical Research Letters*, vol. 29, no. 23, 2002.

Baker, V.R., 'Geological history of water on Mars'. Presentation at Sixth International Conference on Mars, Pasadena, 2003.

Baker, V.R., 'Very recent hydroclimatic change on Mars'. Presentation at 34th Lunar and Planetary Science Conference, Houston, 2003.

Baker, V.R. *et al.*, 'Ancient oceans, ice sheets and the hydrological cycle on

Mars'. *Nature*, vol. 352, 1991, pp. 589–94.

Ball, P., 'Rising damp on the Red Planet'. *Nature Science Update*, 25 January 2001.

Bandfield, J.L. *et al.*, 'Spectroscopic identification of carbonate minerals in the Martian dust.', *Science*, vol. 301, no. 5636, pp. 1084–7, 22 August 2003.

Boynton, W.V. *et al.*, 'Distribution of hydrogen in the near surface of Mars: evidence for subsurface ice deposits'. *Science*, vol. 297, no. 5578, July 2002, pp. 81–5. (Article originally published online 30 May 2002.)

Brandenburg, J.E., 'The New Mars Synthesis: a new concept of Mars' geochemical history'. Presentation at Sixth International Conference on Mars, Pasadena, 2003.

Byrne, S. and Ingersoll. A., 'Martian climatic events inferred from South Polar geomorphology on timescales of centuries'. Presentation at 34th Lunar and Planetary Science Conference, Houston, 2003.

Carr, M and Head, J., 'Oceans on Mars: an assessment of the observational evidence and possible fate'. *Journal of Geophysical Research*, vol. 108, no. E5, 2003.

Christensen, P. *et al.*, 'Water on Mars: evidence from minerals and morphology'. Presentation at Sixth International Conference on Mars, Pasadena, July 2003.

Christensen, P.R., 'Formation of recent Martian gullies through melting of extensive water-rich snow deposits'. *Nature* (advance on-line publication), 19 February 2003.

Clifford, S.M., 'Where is Mars hiding its water?'. *Sky & Telescope*, vol. 106, no. 2, August 2003.

Hecht, M.H., 'Metastability of liquid water on Mars'. Presentation at 32nd Lunar and Planetary Science Conference, Houston, 2001

Hecht. M.H; Bridges, N.T. 'A mechanism for recent production of liquid water on Mars'. Presentation at 34th Lunar and Planetary Science Conference, Houston, 2003.

Kasting, J.F., 'Early Mars: warm enough to melt water?'. Presentation at

AAAS Annual Meeting, Denver, February 2003.

Kerr, R., 'Making a splash with a hint of Mars water'. *Science*, vol. 288, June 2000, pp. 2330–5.

Kerr, R.A., 'Eons of a cold, dry, dusty Mars'. *Science*, vol. 301, no. 5636, 22 August 2003.

Kieffer, H.H., 'Behavior of solid CO_2 on Mars: a real zoo'. Presentation at Sixth International Conference on Mars, Pasadena, 2003.

Kring, D.A. *et al.* 'Cataclysmic bombardment throughout the inner Solar System 3.9–4.0Ga'. *Journal of Geophysical Research*, vol. 107, no. E2, 2002.

Malin, M.C. and Edgett, K.S., 'Sedimentary rocks of early Mars'. *Science*, vol. 290, December 2000, pp. 1927–37.

Malin, M.C and Edgett, K.S. 'Evidence for recent groundwater seepage and surface runoff on Mars'. *Science*, vol. 288, June 2000, pp. 2330–5.

Milstein, M., 'Martian gullies tempt NASA to look for water'. *Nature*, vol. 405, June 2000, p. 987.

Motazedian, T., 'Currently flowing water on Mars'. Paper presented at 34th Lunar and Planetary Science Conference, Houston, 2003.

O'Hanlon, L., 'The outrageous hypothesis'. *Nature*, vol. 413, 2001, pp. 664–6.

Salamunicar, G. *et al.*, 'Intriguing dark streaks on Mars: can we use them for formal proof that we are near the end of large climate change on Mars?' Sixth International Conference on Mars, 2003.

Tanaka, K., 'Fountains of youth'. *Science*, vol. 288, June 2000, p. 2325.

Thompson, J. *et al.*, 'Martian gullies and the stability of water in the Martian environment'. Presentation at 34th Lunar and Planetary Science Conference, Houston, 2003.

Titus, T.N. and Kieffer, H.H., 'Temporaral and spatial distribution of seasonal CO_2 snow and ice'. Presentation at Sixth International Conference on Mars, Pasadena, 2003.

Life on Mars

Ball, P., 'How to spot a Martian'. *Nature* science update, 2 March 2000.

Carlotto, M.J., 'Enigmatic landofrms in Cydonia: geospatial anisotropies, bilateral symmetries and their correlations'. Poster presentation at Sixth International Conference on Mars, 2003.

Gibson, E.K and Pillinger, C.T., 'Beagle 2: seeking the signatures of life on Mars'. Presentation at Sixth International Conference on Mars, Pasadena, July 2003.

Gibson, M.S, McKay, M.F. *et al.*, 'Search for past life on Mars: possible relic biogenic activity in Martian meteorite ALH84001'. *Science*, vol. 273, no. 5277, August 1996, pp. 924–30.

Gillon, J., 'Feedback on Gaia'. *Nature*, vol. 406, 2000, pp. 685–6.

Kerr, R., 'Methane means Martians?'. *ScienceNOW*, March 2004.

Lissauer, J.J., 'How common are habitable planets?' *Nature*, vol. 402, supp. December 1999.

Nisbet, E.G and Sleep, N.H., 'The habitat and nature of early life'. *Nature*, vol. 409, February 2001, pp. 1083–91.

'Search for Martians "too risky" if there's any sign of life'. News in brief article, *Nature*, vol. 417, May 2002, p. 110.

Treiman, A.H. *et al.*, 'Hydrothermal origin for carbonate globules in Martian meteorite ALH84001'. *Earth and Planetary Science Letters*, vol. 204, 2002, pp. 323–32.

Weiss, B. P. *et al.*, 'Atmospheric energy for subsurface life on Mars?'. *PNAS*, vol. 97, 2000, pp. 1395–9.

Exploration of Mars

Baxter, S., 'How NASA lost the case for Mars in 1969'. *Spaceflight, The Journal of the British Interplanetary Society*, vol. 38, June 1996.

Chicarro, A.F., 'The Mars Express mission and its Beagle 2 lander'. Presentation at the Sixth International Conference on Mars, Pasadena, 2003.

'Do we still need astronauts?'. Editorial, *Nature*, vol. 419, October 2002, p. 653.

Dooling, D., 'Nuclear power: the future of spaceflight?' *Space.com*, 22 July 2000.

'Earth invades Mars'. Special report, *Astronomy*, vol. 32, no. 4, April 2004.

Fogg, M.J., 'Terraforming Mars: a review of research'. Published online at www.users.globalnet.co.uk/~mfogg/paper1.htm

Garvin. J.B *et al.*, 'NASA's Mars Exploration Program: scientific strategy 1996–2020'. Sixth International Conference on Mars, 2003.

Golombek, M. *et al.*, 'Mars Exploration Rover landing site selection'. Sixth International Conference on Mars, 2003.

Head, J.W. *et al.*, 'Recent ice ages on Mars'. *Nature*, vol. 426, December 2003, pp. 797–802.

Hiscox, J.A., 'Biology and the planetary engineering of Mars'. Paper published on the *Case for Mars* website, http://spot.colorado.edu/~marscase/cfm/articles/biorev3.html

Jakosky, B., 'The future of Mars exploration'. Presentation at AAAS Annual Meeting, Denver, February 2003.

'Mars in focus'. Special feature, *Spaceflight, The Journal of the British Interplanetary Society*, vol. 46, no. 1, January 2004.

'Mars reveals its secrets'. Special report, *Astronomy Now*, April 2004.

'Mars'. Nature Insight special, *Nature*, vol. 412, July 2001, pp. 209–57.

'Martian landings'. Special feature, *Astronomy Now*, January 2004.

'MER landing sites'. Presentation at Sixth International Conference on Mars, Pasadena, 2003.

'Mission status reports [Beagle 2]'. Lander Operations Control Centre, University of Leicester/Beagle 2, December 2003, January–February 2004.

Mullins, J., 'Sink or swim'. *New Scientist*, vol. 178, no. 2396, 24 May 2003.

Portree, D.S.F., 'Closer to Mars'. *Air & Space*, October/ November 2001.

Portree, D.S.F., 'Humans to Mars: fifty years of mission planning 1950–2000'. *Monographs in Aerospace History*, no. 21, NASA SP-2001-4521, History Division, National Aeronautics and Space Administration, February 2001.

Portree, D.S.F., 'NASA's Mars Design Reference Mission: it originated with Mars Direct – but where did Mars Direct come from?' *Space Times*, November–December 1999.

Portree, D.S.F. 'The New Martian Chronicles'. *Astronomy*, August 1997.

'Report of the 90-Day Study on Human Exploration of the Moon and Mars'. Cost Summary, unpublished chapter, Washington, DC: NASA, November 1989

'Return to Mars'. Special feature, *The Planetary Report*, vol. XXIII, no. 3, May/June 2003.

'The Beagle 2 Bulletin'. Regular newssheet produced by the Beagle 2 team.

WEBSITES

Romance to Reality – David Portree's comprehensive history of what might have been:
http://members.aol.com/dsfportree/explore.htm

Nick Hoffman's White Mars pages:
www.earthsci.unimelb.edu.au/mars/Enter.html

Mars in ancient and popular culture:
http://humbabe.arc.nasa.gov/mgcm/fun/ancient_mars.html

The Enterprise Mission – Richard Hoagland's musings on the 'face' on Mars, and other controversies:
www.enterprisemission.com

NASA's gateway site:
www.nasa.gov

The European Space Agency:
www.esa.int

The Jet Propulsion Laboratory:
www.jpl.nasa.gov

A history of NASA robotic space exploration:
http://spacescience.nasa.gov/missions/index.htm

The 2004 Mars Rovers:
http://marsrovers.nasa.gov/home/index.html

Mars Express:
http://www.esa.int/SPECIALS/Mars_Express/index.html

Beagle 2:
http://www.beagle2.com/index.htm

The Planetary Society:
www.planetary.org

Some general space news sites:
www.space.com
www.spacedaily.com
www.marsdaily.com

Index

Note: Italics are used to refer to illustrations